Garrison

KUMON MATH WORKBOOKS

Grade **5**

W9-AMM-286

Geometry & Measurement

Table of Contents

KUM⊙N

Review

1 If you round the number below to the nearest thousand, it would be 90,000. Write the numbers that could possibly fit in the box below.   8 points

$$90,\boxed{}50$$

(   )

2 Write the appropriate number in each box below.   6 points per question

(1) 3.2 is the number you get from adding $\boxed{}$ parts of 0.1.

(2) The number you get from adding 59 parts of 0.1 is $\boxed{}$.

(3) The number you get from adding 48 parts of 0.1 is $\boxed{}$.

3 Convert each improper fraction into a mixed fraction or an integer.   5 points per question

(1) $\dfrac{9}{7} = \boxed{}$ (2) $\dfrac{15}{5} = \boxed{}$

4 Find the area of the shapes below.   6 points per question

(1)

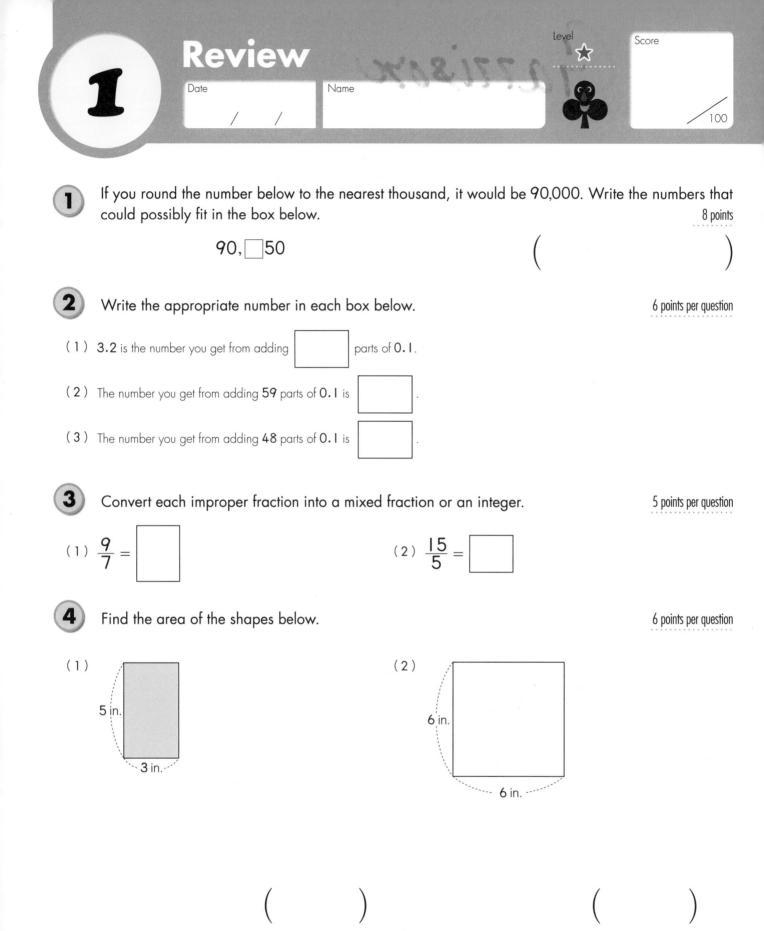

5 in.
3 in.

(2)

6 in.
6 in.

(   )   (   )

5 There are two similar circles inside the larger circle on the right. The larger circle has a diameter of 10 centimeters.

6 points per question

(1) What is the radius of the big circle?   (   )

(2) What is the diameter of each of the small circles?   (   )

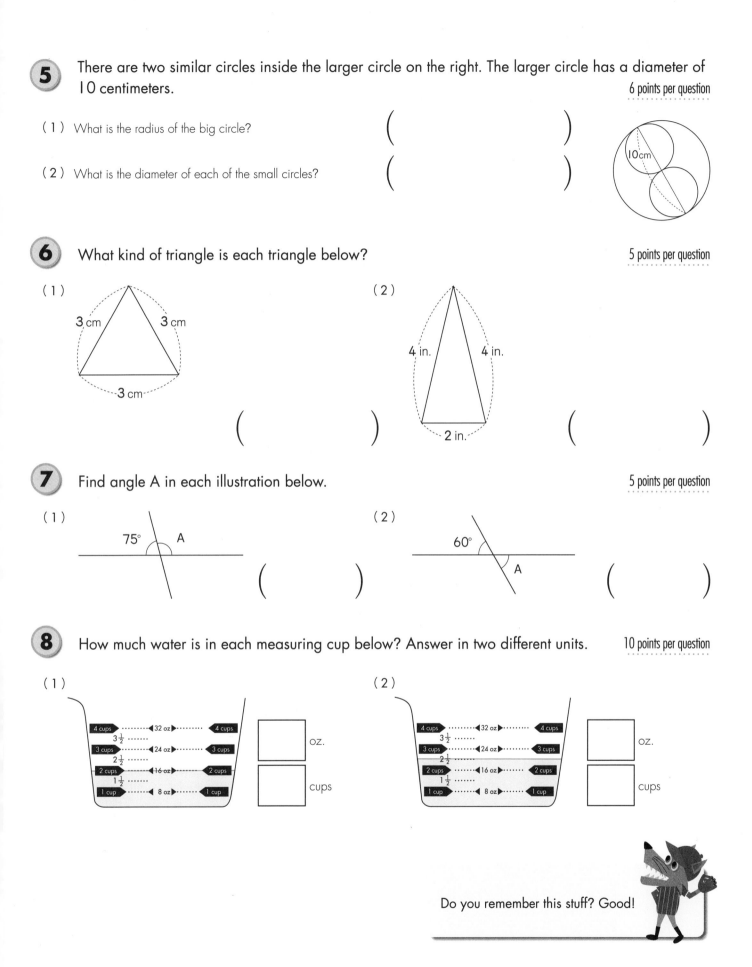

6 What kind of triangle is each triangle below?

5 points per question

(1)

3 cm   3 cm

3 cm

(   )

(2)

4 in.   4 in.

2 in.

(   )

7 Find angle A in each illustration below.

5 points per question

(1)

75°   A

(   )

(2)

60°

A

(   )

8 How much water is in each measuring cup below? Answer in two different units.

10 points per question

(1)

4 cups ······ ◄32 oz► ······ 4 cups
3½ ·······
3 cups ······ ◄24 oz► ······ 3 cups
2½ ·······
2 cups ······ ◄16 oz► ······ 2 cups
1½ ·······
1 cup ······ ◄ 8 oz► ······ 1 cup

oz.

cups

(2)

4 cups ······ ◄32 oz► ······ 4 cups
3½ ·······
3 cups ······ ◄24 oz► ······ 3 cups
2½ ·······
2 cups ······ ◄16 oz► ······ 2 cups
1½ ·······
1 cup ······ ◄ 8 oz► ······ 1 cup

oz.

cups

Do you remember this stuff? Good!

Review

2

Date　/　/

Name

1 The number below has been rounded to the nearest hundred. What is the possible range for the original number?　　6 points

4,500

From (　　　) to (　　　)

2 Write the appropriate number in each box below.　　5 points per question

(1) $\frac{3}{4}$ is ☐ parts of $\frac{1}{4}$.

(2) 5 parts of ☐ is $\frac{5}{8}$.

(3) 5 parts of $\frac{1}{5}$ is ☐, which is equal to ☐.

3 Circle the larger number.　　5 points per question

(1) [0.3　　0.7]

(2) [5.2　　4]

(3) [1　　9.9]

(4) [26　　6.6]

4 Find the area of the shape below.　　7 points

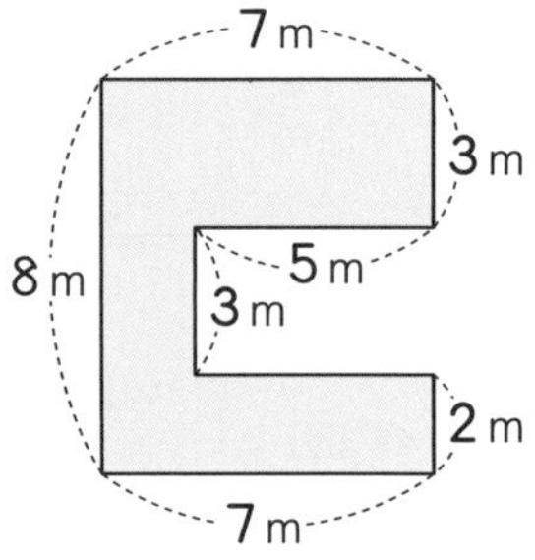

7 m
3 m
8 m
5 m
3 m
2 m
7 m

(　　　)

5 Convert each mixed fraction into an improper fraction.　　4 points per question

(1) $1\frac{1}{3}$ = ☐

(2) $2\frac{4}{5}$ = ☐

6 As pictured on the right, you have eight similar balls that fit snugly inside one box that is 6 inches long.

5 points per question

(1) What is the diameter of each ball?

(   )

(2) What is the width of the box?

(   )

7 Draw a triangle with sides of 3 centimeters, 3 centimeters and 4 centimeters.

6 points

8 Use your protractor to measure each angle below.

5 points per question

(1)

(   )

(2)

A

(   )

9 Read each clock below. Then add the amount of time in brackets to find the final time. 6 points per question

(1) (A.M.)

[2 hours later]

(   )

(2) (P.M.)

[An hour and a half later]

(   )

(3) (A.M.)

[2 hours and 30 minutes later]

(   )

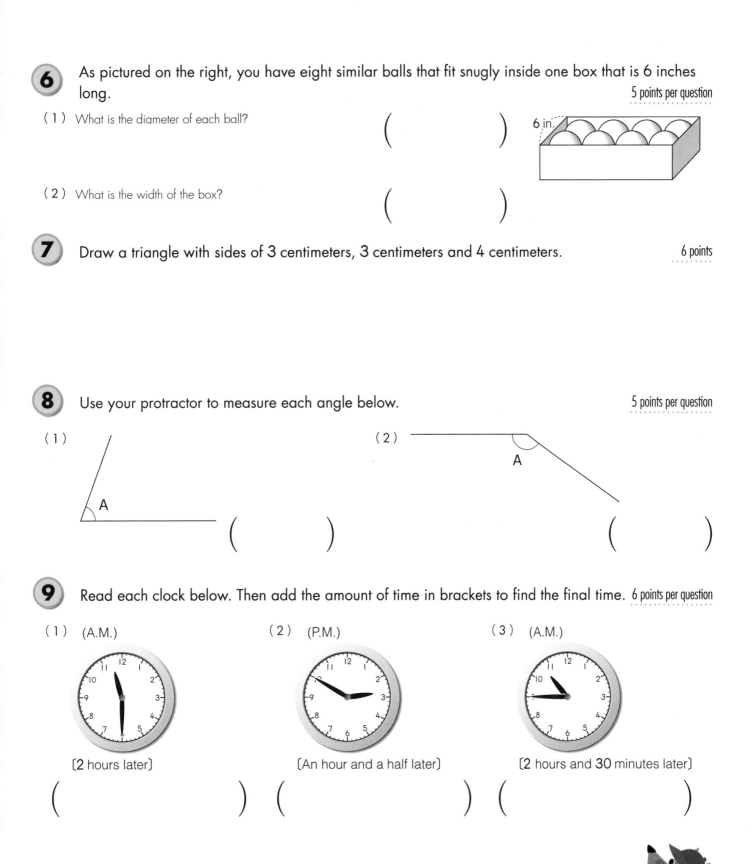

Let's get this party started!

3 Odd & Even Numbers ★★

Level

Score

/100

Date   /   /

Name

1 Divide whole numbers from 0 through 10 into numbers than can be divided by 2 and numbers that cannot be divided by 2. Write the appropriate numbers in the boxes below.

5 points per question

(1) Numbers that can be divided by 2 ⋯⋯ 0 , 2 , 4 , ☐ , ☐ , ☐

(2) Numbers that can't be divided by 2 ⋯⋯ 1 , 3 , 5 , ☐ , ☐

> **Don't forget!**
>
> Numbers with a 0, 2, 4, 6, or 8 in the ones place are **even** numbers. Numbers with a 1, 3, 5, 7, or 9 in the ones place are **odd** numbers.

2 Label each number below as odd or even.

3 points per question

(1) 1   (   )   (2) 0   (   )

(3) 4   (   )   (4) 3   (   )

(5) 7   (   )   (6) 9   (   )

(7) 5   (   )   (8) 8   (   )

(9) 6   (   )   (10) 10   (   )

(11) 11   (   )   (12) 12   (   )

3 Sort the numbers from 11 to 30 into odd and even numbers below.

4 points per question

(1) Write all the even numbers.   (   )

(2) Write all the odd numbers.   (   )

> **Don't forget!**
>
> Numbers that can be divided by 2 are called **even** numbers (including 0).
> Numbers that cannot be divided by 2 are called **odd** numbers.

6   © Kumon Publishing Co., Ltd.

4 Circle the even numbers below.   10 points

46, 53, 68, 71, 87, 96, 104, 113, 126,
135, 147, 156, 160, 177, 189, 196

5 Circle the odd numbers below.   10 points

43, 58, 69, 76, 84, 91, 108, 117, 128,
134, 145, 151, 166, 179, 182, 199

6 Sort the numbers below into even and odd.   5 points per question

36, 43, 126, 180, 193, 252, 300, 78,
83, 67, 98, 112, 136, 145, 321,
267, 284, 176, 184, 365, 243

Even (   )

Odd (   )

7 You have cards with the numbers 1, 2 and 3 on them. Use the cards to answer the questions below.   5 points per question

(1) Write two even numbers you can make with the cards.   (   )(   )

(2) Write four odd numbers you can make with the cards.   (   )(   )(   )(   )

8 Write the largest odd number you can make with the numbers 2, 3 and 4.   6 points

(   )

Okay, let's switch it up a little.
You're doing great!

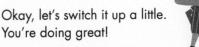

4 Decimals

Date   /   /

Name

Level ☆☆

Score   / 100

1 How many parts of 0.1 are in each number below?   2 points per question

(1) 0.7 (7) (2) 0.9 (   ) (3) 1 (10) (4) 1.1 (   )

2 How many parts of 0.01 are in each number below?   2 points per question

(1) 0.04 (4)   (2) 0.08 (   )   (3) 0.02 (   )

(4) 0.03 (   )   (5) 0.1 (10)   (6) 0.11 (   )

3 How many parts of 0.001 are in each number below?   2 points per question

(1) 0.006 (6)   (2) 0.005 (   )   (3) 0.002 (   )

(4) 0.003 (   )   (5) 0.008 (   )   (6) 0.01 (10)

4 Write the appropriate numbers below.   2 points per question

(1) Add 3 parts of 0.1   (0.3)   (2) Add 7 parts of 0.1   (   )

(3) Add 15 parts of 0.1   (   )   (4) Add 48 parts of 0.1   (   )

5 Write the appropriate numbers below.   2 points per question

(1) Add 5 parts of 0.01   (0.05)   (2) Add 9 parts of 0.01   (   )

(3) Add 15 parts of 0.01 (   )   (4) Add 125 parts of 0.01. (   )

6 Write the appropriate numbers below.   2 points per question

(1) Add 4 parts of 0.001 (0.004)   (2) Add 18 parts of 0.001 (   )

(3) Add 392 parts of 0.001 (   )   (4) Add 1,549 parts of 0.001 (   )

Don't forget!

In order to get the number 2.734, you have to add 2 parts of 1, 7 parts of 0.1, 3 parts of 0.01, and 4 parts of 0.001.

2	.	7	3	4
ones place		tenths place	hundredths place	thousandths place

7 Answer the following questions about the number 3.456. 4 points per question

(1) 3 in the ones place means that you've added ☐ parts of 1.

(2) 4 in the tenths place means that you've added ☐ parts of 0.1.

(3) 5 in the hundredths place means that you've added ☐ parts of 0.01.

(4) 6 in the thousandths place means that you've added ☐ parts of 0.001.

8 How many parts of 0.1, 0.01 and 0.001 are in each number below? 4 points per question

(1) The number of 3.512 has ☐ parts of 1 and ☐ parts of 0.1, ☐ part of 0.01, ☐ parts of 0.001.

(2) The number of 0.275 has ☐ parts of 1 and ☐ parts of 0.1, ☐ parts of 0.01, ☐ parts of 0.001.

(3) The number of 0.508 has ☐ parts of 1 and ☐ parts of 0.1, ☐ parts of 0.01, ☐ parts of 0.001.

9 Write the appropriate numbers below. 4 points per question

(1) The number you get from adding 2 parts of 1, 3 parts of 0.1 and 5 parts of 0.01. ( )

(2) The number you get from adding 3 parts of 1, 4 parts of 0.1, 1 part of 0.01, and 7 parts of 0.001. ( )

(3) The number you get from adding 2 parts of 1, 7 parts of 0.1 and 4 parts of 0.001. ( )

(4) The number you get from adding 8 parts of 0.1, 6 parts of 0.01, and 4 parts of 0.001. ( )

No worries, right? Good job.

Decimals

Level ☆☆

Date / /

Name

Score /100

1 Write the appropriate number in each box on the number lines below. *2 points per box*

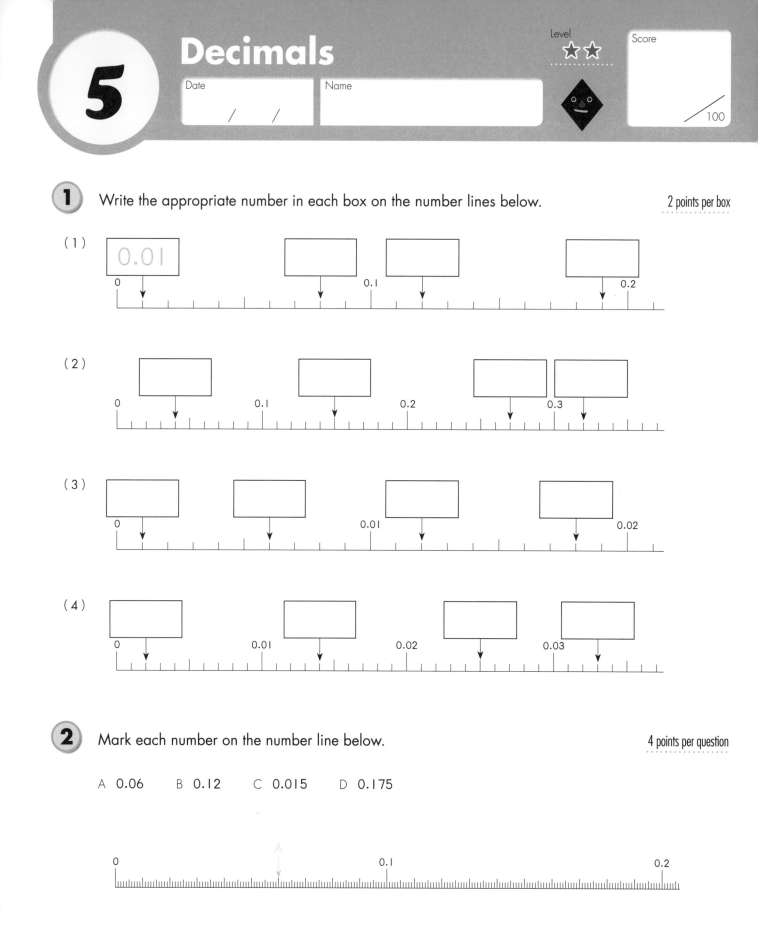

(1)

| 0.01 |

0   0.1   0.2

(2)

0   0.1   0.2   0.3

(3)

0   0.01   0.02

(4)

0   0.01   0.02   0.03

2 Mark each number on the number line below. *4 points per question*

A 0.06   B 0.12   C 0.015   D 0.175

0   0.1   0.2

3 Write a ✓ under the larger number in each pair below.

4 points per question

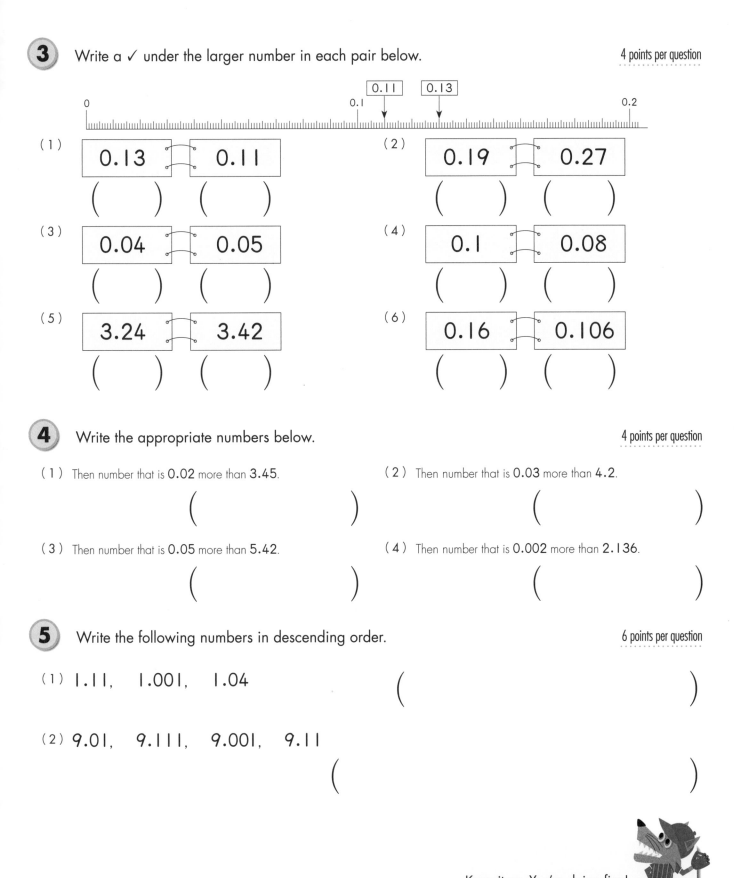

(1) 0.13 ⟷ 0.11
( ) ( )

(2) 0.19 ⟷ 0.27
( ) ( )

(3) 0.04 ⟷ 0.05
( ) ( )

(4) 0.1 ⟷ 0.08
( ) ( )

(5) 3.24 ⟷ 3.42
( ) ( )

(6) 0.16 ⟷ 0.106
( ) ( )

4 Write the appropriate numbers below.

4 points per question

(1) Then number that is **0.02** more than **3.45**.

( )

(2) Then number that is **0.03** more than **4.2**.

( )

(3) Then number that is **0.05** more than **5.42**.

( )

(4) Then number that is **0.002** more than **2.136**.

( )

5 Write the following numbers in descending order.

6 points per question

(1) 1.11, 1.001, 1.04

( )

(2) 9.01, 9.111, 9.001, 9.11

( )

Keep it up. You're doing fine!

6 Decimals

Date / /   Name

Score /100

---

1 Write the appropriate number in each box below.

4 points per question

(1) The number that is 10 times 2 is the same as adding 10 parts of 2, and is ⬚ . This is shown by the number

sentence: $2 \times 10 =$ ⬚

(2) The number that is 10 times 0.2 is the same as adding 10 parts of 0.2, and is ⬚ . This is shown by the number

sentence: $0.2 \times 10 =$ ⬚

(3) The number that is 10 times 0.02 is the same as adding 10 parts of 0.02, and is ⬚ . This is shown by the number

sentence: $0.02 \times 10 =$ ⬚

---

2 Write the appropriate number in each box below.

3 points per question

(1) The number that is 100 times 0.2 is the same as adding 100 parts of 0.2 and is ⬚ . This is shown by the number

sentence: $0.2 \times 100 =$ ⬚

(2) The number that is 100 times 0.02 is the same as adding 100 parts of 0.02 and is ⬚ . This is shown by the number

sentence: $0.02 \times 100 =$ ⬚

---

3 Write the appropriate numbers below.

4 points per question

(1) 10 times 0.3   (   )   (2) 10 times 0.03   (   )

(3) 10 times 0.9   (   )   (4) 10 times 0.05   (   )

(5) 10 times 0.4   (   )   (6) 10 times 0.04   (   )

(7) 10 times 0.7   (   )   (8) 10 times 0.08   (   )

© Kumon Publishing Co., Ltd.

4 Write the appropriate number in each box below.

4 points per question

(1) The number that is one tenth of 20 is the same as dividing 20 into 10 parts, and is ☐ . This is shown by the number

sentence: $20 \div 10 =$ ☐

(2) The number that is one tenth of 2 is the same as dividing 2 into 10 parts, and is ☐ . This is shown by the number

sentence: $2 \div 10 =$ ☐

(3) The number that is one tenth of 0.2 is the same as dividing 0.2 into 10 parts, and is ☐ . This is shown by the number

sentence: $0.2 \div 10 =$ ☐

5 Write the appropriate number in each box below.

3 points per question

(1) The number that is one hundredth of 2 is the same as dividing 2 into 100 parts, and is ☐ . This is shown by the number

sentence: $2 \div 100 =$ ☐

(2) The number that is one hundredth of 0.2 is the same as dividing 0.2 into 100 parts, and is ☐ . This is shown by the

number sentence: $0.2 \div 100 =$ ☐

6 Write the appropriate numbers below.

4 points per question

(1) One tenth of 4 ( )   (2) One tenth of 0.4 ( )

(3) One tenth of 7 ( )   (4) One tenth of 0.6 ( )

(5) One hundredth of 3 ( )   (6) One hundredth of 0.3 ( )

(7) One hundredth of 8 ( )   (8) One hundredth of 0.5 ( )

You've got it. Excellent!

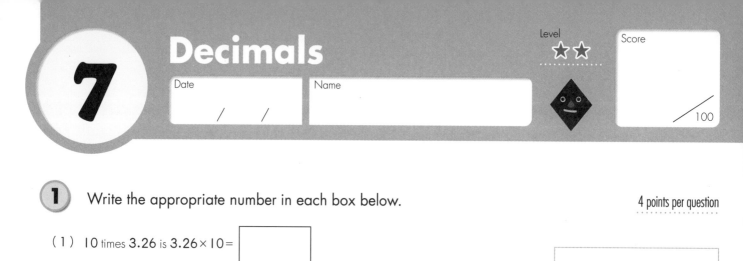

7 Decimals

Date / /

Name

1 Write the appropriate number in each box below.

4 points per question

(1) 10 times **3.26** is **3.26 × 10 =** ☐

When you multiply a number by **10**, the decimal point moves ☐ digit(s) to the right.

(2) 100 times **3.26** is **3.26 × 100 =** ☐

When you multiply a number by **100**, the decimal point moves ☐ digit(s) to the right.

original number	3.26
ten times	32.6
hundred times	326.0

Don't forget!

When you multiply a number by 10, the decimal point moves 1 digit to the right. When you multiply a number by 100, the decimal point moves 2 digits to the right.

2 Write the appropriate numbers below.

4 points per question

(1) 10 times **0.4**   (   )   (2) 10 times **4.2**   (   )

(3) 10 times **2.57**   (   )   (4) 10 times **46.32**   (   )

(5) 100 times **0.7**   (   )   (6) 100 times **0.32**   (   )

(7) 100 times **2.47**   (   )   (8) 100 times **18.24**   (   )

(9) 100 times **3.06**   (   )   (10) 100 times **41.05**   (   )

3 How many times the left number is the right number?

2 points per question

(1) **32.8 → 328**   (   ) times   (2) **5.36 → 536**   (   ) times

© Kumon Publishing Co., Ltd.

4 Write the appropriate number in each box below.

4 points per question

(1) A tenth of **32.6** is **32.6 ÷ 10** = ☐

When you divide a number by **10**, the decimal point moves ☐ digit(s) to the left.

(2) A hundredth of **32.6** is **32.6 ÷ 100** = ☐

When you divide a number by **100**, the decimal point moves ☐ digit(s) to the left.

original number	32.6
one tenth	3.26
one hundredth	0.326

Don't forget!

When you divide a number by 10, the decimal point moves 1 digit to the left. When you divide a number by 100, the decimal point moves 2 digits to the left.

5 Write the appropriate numbers below.

3 points per question

(1) One tenth of **32**   (   )   (2) One tenth of **2.4**   (   )

(3) One tenth of **2.57**   (   )   (4) One tenth of **48.7**   (   )

(5) One hundredth of **463**   (   )   (6) One hundredth of **230**   (   )

(7) One hundredth of **15.8**   (   )   (8) One hundredth of **7.5**   (   )

6 What fraction of the left number is the right number?

3 points per question

(1) 23.8 → 2.38   (   )   (2) 456 → 4.56   (   )

7 Write the numbers that are 10 times, 100 times, one tenth, and one hundredth of the number 326.5.

10 points for completion

ten times	hundred times	one tenth	one hundredth
( )	( )	( )	( )

Phew! Take a break if you need one.

15

Decimals & Measurements

Level ☆☆

Date / /

Name

Score /100

1 100 centimeters = 1 meter. Use decimals to answer the questions below.

2 points per question

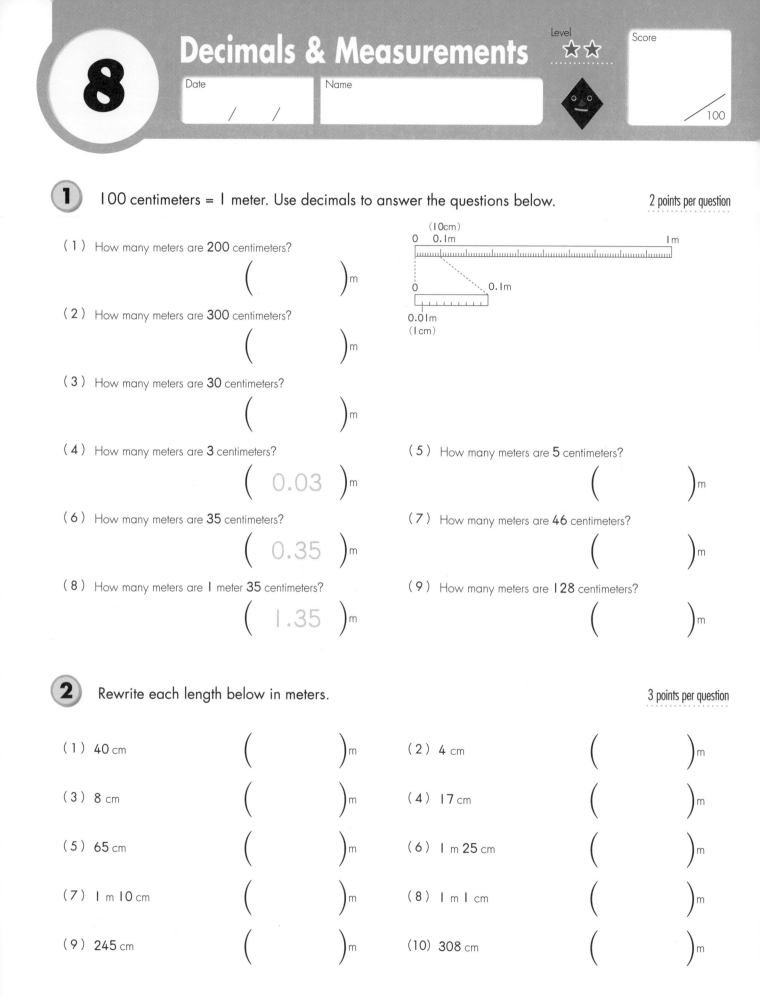

(1) How many meters are **200** centimeters?

( )m

(2) How many meters are **300** centimeters?

( )m

(3) How many meters are **30** centimeters?

( )m

(4) How many meters are **3** centimeters?

(0.03)m

(5) How many meters are **5** centimeters?

( )m

(6) How many meters are **35** centimeters?

(0.35)m

(7) How many meters are **46** centimeters?

( )m

(8) How many meters are **1** meter **35** centimeters?

(1.35)m

(9) How many meters are **128** centimeters?

( )m

2 Rewrite each length below in meters.

3 points per question

(1) 40 cm ( )m

(2) 4 cm ( )m

(3) 8 cm ( )m

(4) 17 cm ( )m

(5) 65 cm ( )m

(6) 1 m 25 cm ( )m

(7) 1 m 10 cm ( )m

(8) 1 m 1 cm ( )m

(9) 245 cm ( )m

(10) 308 cm ( )m

3 1,000 meters = 1 kilometer. Use decimals to answer the questions below.

3 points per question

(1) How many kilometers are 3,000 meters?

(　　　　　)km

(100m)
0　0.1km　　　　　　　　　　　　　　　　1 km

0　　　　0.1km

0　　　　0.01 km(10 m)

0.001km
(1m)

(2) How many kilometers are 300 meters?

(　　　　　)km

(3) How many kilometers are 30 meters?

(　　　　　)km

(4) How many kilometers are 3 meters?

(0.003)km

(5) How many kilometers are 354 meters?

(　　　　　)km

(6) How many kilometers are 65 meters?

(　　　　　)km

(7) How many kilometers are 1 kilometer 450 meters?

(　　　　　)km

(8) How many kilometers are 2,387 meters?

(　　　　　)km

4 Rewrite each length below in kilometers.

2 points per question

(1) 600 m （　　　　　）km　　　(2) 150 m （　　　　　）km

(3) 40 m （　　　　　）km　　　(4) 8 m （　　　　　）km

(5) 275 m （　　　　　）km　　　(6) 36 m （　　　　　）km

(7) 1 km 480 m （　　　　　）km　　　(8) 2,306 m （　　　　　）km

5 Rewrite each weight below in kilograms.

2 points per question

(1) 500 g （　　　　　）kg　　　(2) 30 g （　　　　　）kg

(3) 725 g （　　　　　）kg　　　(4) 64 g （　　　　　）kg

(5) 1,820 g （　　　　　）kg　　　(6) 2,905 g （　　　　　）kg

Okay, time to try something new.
Are you ready?

9 Decimals & Fractions

Level ★★

Date / /

Name

Score

/100

1 Write the appropriate fraction in each box on the number lines below.

4 points per box

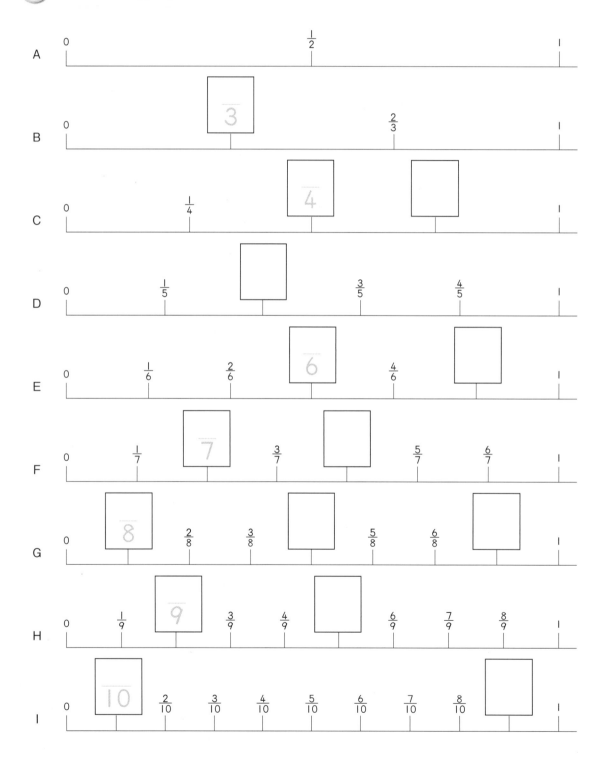

2 Use the number lines to answer the questions below.

4 points per question

(1) Write all the fractions that are on the number lines that are equal to $\frac{1}{2}$.　　（　　　　　　）

(2) Write all the fractions that are on the number lines that are equal to $\frac{1}{3}$.　　（　　　　　　）

(3) Write all the fractions that are on the number lines that are equal to $\frac{6}{9}$.　　（　　　　　　）

3 Write the larger number from each pair of fractions in the brackets.

4 points per question

(1) $\left[\ \frac{1}{4}\quad\frac{3}{4}\ \right]$　　（　　　　）　　　　(2) $\left[\ \frac{2}{3}\quad\frac{2}{5}\ \right]$　　（　　　　）

Don't forget!

When comparing two fractions with the same denominator, the fraction with a larger numerator is the larger number.
When comparing two fractions with the same numerator, the fraction with the smaller denominator is the larger number.

4 Write the following fractions in descending order.

4 points per question

(1) $\frac{4}{5}$, $\frac{1}{5}$, $\frac{2}{5}$, $\frac{3}{5}$　　　　　　（　　　　　　　　）

(2) $\frac{3}{7}$, $\frac{3}{4}$, $\frac{3}{9}$, $\frac{3}{8}$　　　　　　（　　　　　　　　）

5 Write the following fractions in ascending order.

4 points per question

(1) $\frac{5}{8}$, $\frac{2}{8}$, $\frac{11}{8}$, $\frac{9}{8}$　　　　　　（　　　　　　　　）

(2) $\frac{4}{3}$, $\frac{4}{7}$, $\frac{4}{5}$, $\frac{4}{9}$　　　　　　（　　　　　　　　）

(3) $\frac{5}{6}$, $\frac{3}{6}$, $\frac{7}{6}$, $\frac{6}{6}$　　　　　　（　　　　　　　　）

No problems so far, right? Good news.

Decimals & Fractions

10

Level ★★

Date / /

Name

Score /100

Don't forget!

The quotient in division can also be represented in a fraction.   ⟨Example⟩ $3 \div 5 = \frac{3}{5}$

---

1 Rewrite each division problem as a fraction.

1 point per question

(1) $3 \div 4 = \frac{3}{4}$

(2) $5 \div 7 = \frac{\square}{\square}$

(3) $1 \div 3 =$

(4) $7 \div 11 =$

(5) $4 \div 9 =$

(6) $9 \div 10 =$

(7) $7 \div 8 =$

(8) $3 \div 7 =$

(9) $5 \div 12 =$

(10) $9 \div 14 =$

(11) $5 \div 4 = \frac{5}{4}$

(12) $7 \div 3 =$

(13) $10 \div 7 =$

(14) $9 \div 4 =$

(15) $11 \div 6 =$

(16) $6 \div 5 =$

(17) $12 \div 7 =$

(18) $15 \div 8 =$

2 Rewrite each fraction as a division problem.

2 points per question

(1) $\frac{2}{5} = 2 \div \square$

(2) $\frac{1}{6} = \square \div 6$

(3) $\frac{7}{4} = \square \div 4$

(4) $\frac{11}{9} = 11 \div \square$

3 Rewrite each fraction as a decimal.

1 point per question

(1) $\frac{2}{5} =$

(2) $\frac{3}{5} =$

(3) $\frac{4}{5} =$

(4) $\frac{6}{5} =$

(5) $\frac{3}{4} =$

(6) $\frac{5}{4} =$

(7) $\frac{3}{2} =$

(8) $\frac{5}{2} =$

(9) $\frac{7}{8} =$

(10) $\frac{9}{8} =$

(11) $\frac{1}{10} =$

(12) $\frac{17}{10} =$

(13) $\frac{1}{4} =$

(14) $\frac{1}{8} =$

(15) $\frac{1}{16} =$

(16) $\frac{5}{16} =$

20   © Kumon Publishing Co., Ltd.

Don't forget!

$0.1 = \dfrac{1}{10}, \quad 0.2 = \dfrac{2}{10}, \quad 0.3 = \dfrac{3}{10}, \cdots\cdots$

$0.01 = \dfrac{1}{100}, \quad 0.02 = \dfrac{2}{100}, \quad 0.03 = \dfrac{3}{100}, \cdots\cdots$

4 Convert each decimal below into a fraction.   1 point per question

(1) $0.3 = \dfrac{}{10}$

(2) $0.7 =$

(3) $0.9 =$

(4) $0.4 =$

(5) $0.5 =$

(6) $1.1 = \dfrac{}{10}$

(7) $1.3 =$

(8) $1.8 =$

(9) $2.7 =$

(10) $3.9 =$

(11) $0.03 = \dfrac{}{100}$

(12) $0.07 =$

(13) $0.09 =$

(14) $0.08 =$

(15) $0.13 =$

(16) $0.27 =$

(17) $0.35 =$

(18) $1.13 = \dfrac{}{100}$

(19) $2.23 =$

(20) $1.07 =$

5 Convert each decimal into a fraction, and each fraction into a decimal below.   2 points per question

(1) $0.1 =$

(2) $1.5 =$

(3) $0.71 =$

(4) $2.49 =$

(5) $1.01 =$

(6) $\dfrac{9}{10} =$

(7) $\dfrac{4}{5} =$

(8) $\dfrac{6}{4} =$

(9) $\dfrac{7}{20} =$

(10) $\dfrac{37}{100} =$

6 Circle the larger number in each pair of numbers below.   2 points per question

(1) $\left[\ \dfrac{1}{2} \quad 0.4 \ \right]$

(2) $\left[\ 0.25 \quad \dfrac{3}{10} \ \right]$

(3) $\left[\ 1.5 \quad \dfrac{8}{5} \ \right]$

(4) $\left[\ \dfrac{36}{100} \quad 0.35 \ \right]$

7 Write the following numbers in descending order.   10 points for completion

$\left[\ 1.6, \ 0.7, \ \dfrac{3}{4}, \ \dfrac{9}{5}, \ 1\dfrac{1}{2} \ \right]$ $\Big(\hspace{6cm} \Big)$

Let's do some geometry! Ready?

11

Perpendicular & Parallel Lines ★★

Level

Date   /   /

Name

Score

/100

1 Name all the right angles among the angles below.

20 points for completion

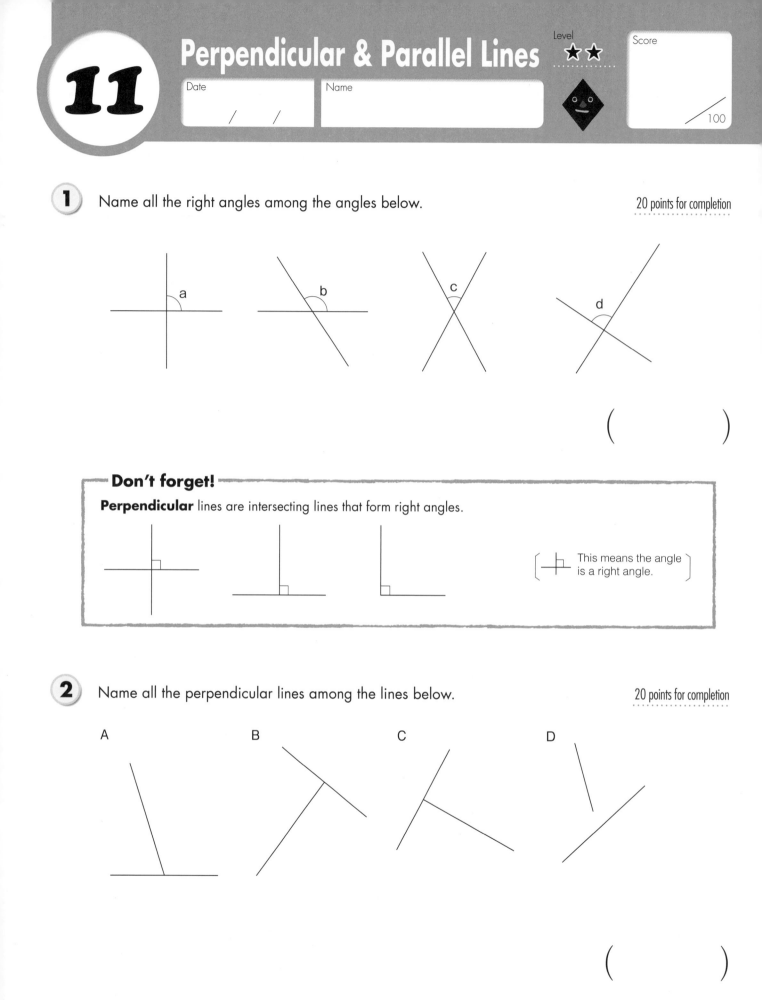

(   )

Don't forget!

Perpendicular lines are intersecting lines that form right angles.

This means the angle is a right angle.

---

2 Name all the perpendicular lines among the lines below.

20 points for completion

A   B   C   D

(   )

3 Which lines are perpendicular to line A?

20 points for completion

( )

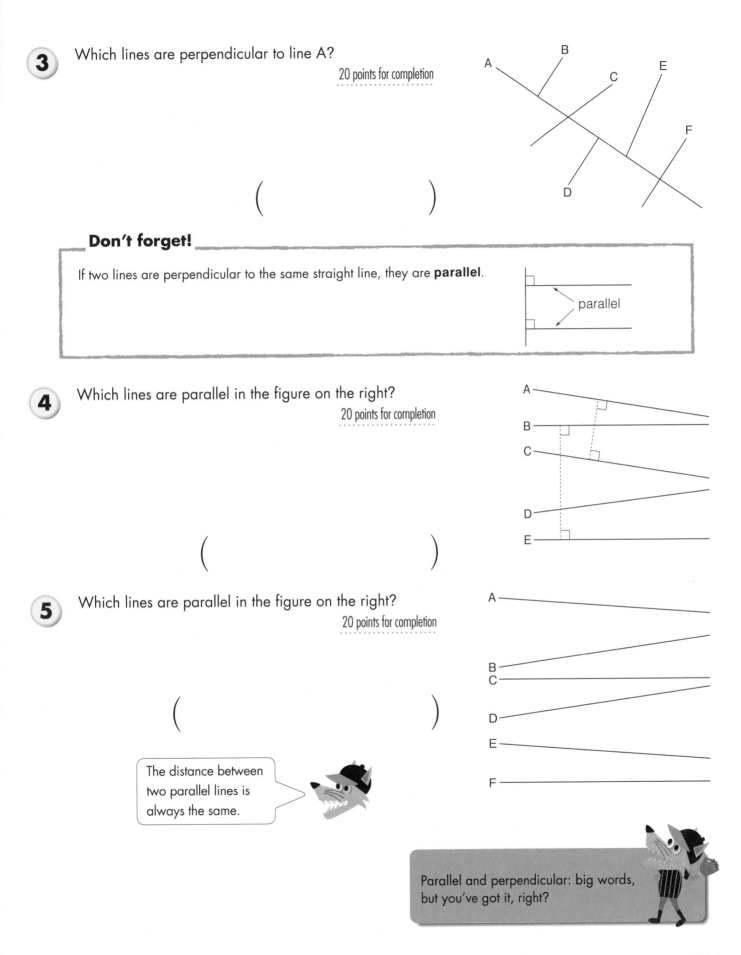

Don't forget!

If two lines are perpendicular to the same straight line, they are **parallel**.

parallel

4 Which lines are parallel in the figure on the right?

20 points for completion

( )

5 Which lines are parallel in the figure on the right?

20 points for completion

( )

The distance between two parallel lines is always the same.

Parallel and perpendicular: big words, but you've got it, right?

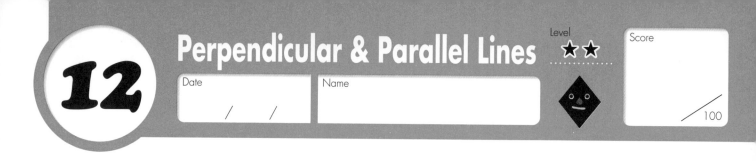

Perpendicular & Parallel Lines

Level ★★

Date   /   /

Name

Score   /100

1 The lines A, B and C are parallel in the figure to the right. If a straight line intersects two parallel lines at an angle, that angle will be the same for both parallel lines.   6 points per question

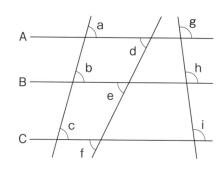

(1) Which angles are equal to **a**?

(b,   )

(2) Which angles are equal to **d**?

(   )

(3) Which angles are equal to **g**?

(   )

2 Line C intersects the parallel lines A and B as shown in the figure on the right. Write the two different groups of equal angles.   8 points for completion

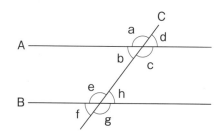

(   ) , (   )

3 Line C intersects the parallel lines A and B as shown in the figure on the right. Use the figure to answer the questions below.   5 points per question

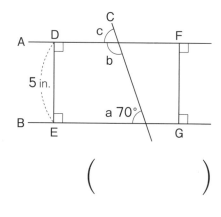

(1) How long is line **FG**?

(   )

(2) Find angle **c**.

(   )

(3) Find angle **b**.

(   )

(4) What is the sum of angles **a** and **b**?

(   )

4 Lines A and B are parallel, and angle a is 120° in the figure on the right.

6 points per question

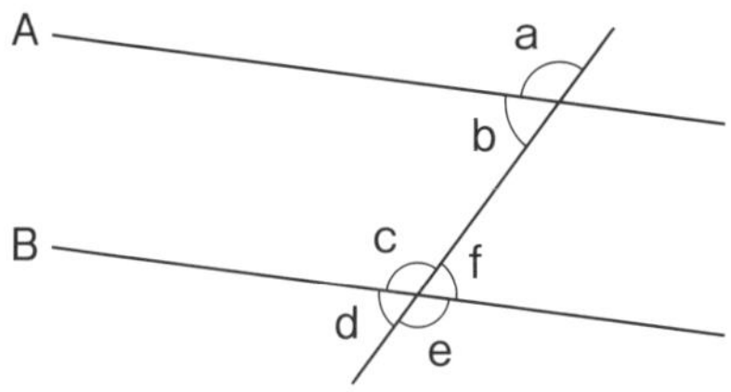

(1) Which angles are equal to angle **a**?

(   )

(2) Find angle **b**.

(   )

(3) Which angles are equal to angle **b**?

(   )

(4) What is the sum of angles **b** and **c**?

(   )

5 Lines A and B are parallel, and lines C and D are parallel.

6 points per question

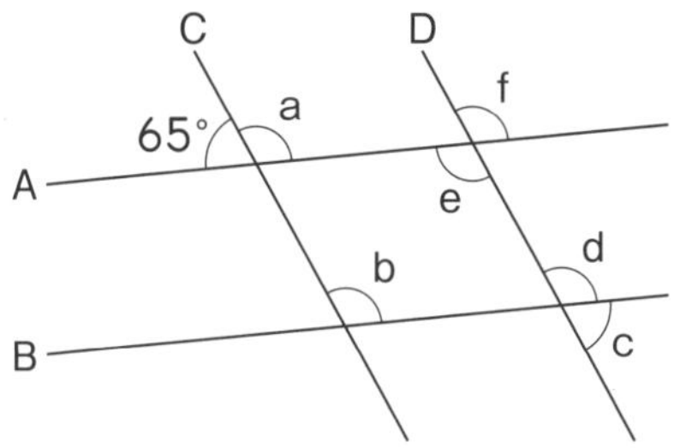

(1) Which angles are equal to angle **b**?

(   )

(2) Find angle **b**.

(   )

(3) Which angles are 65°?

(   )

6 Answer the questions below about the shapes pictured here.

6 points per question

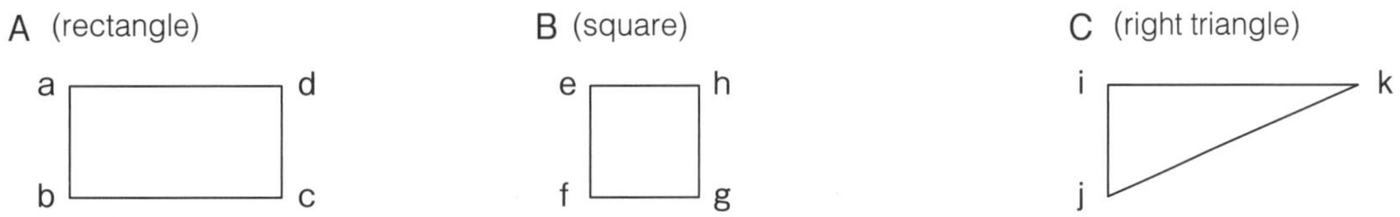

A (rectangle)

B (square)

C (right triangle)

(1) Mark all the perpendicular intersections.

(2) Write all the pairs of parallel lines in the shapes above.

(   )

You're doing great. Keep it up!

   25

Date   /   /

Name

Score

/100

1 Using the example as a guide, use your triangular ruler to draw a line that is perpendicular to each line A below. The second line should go through point B.

9 points per question

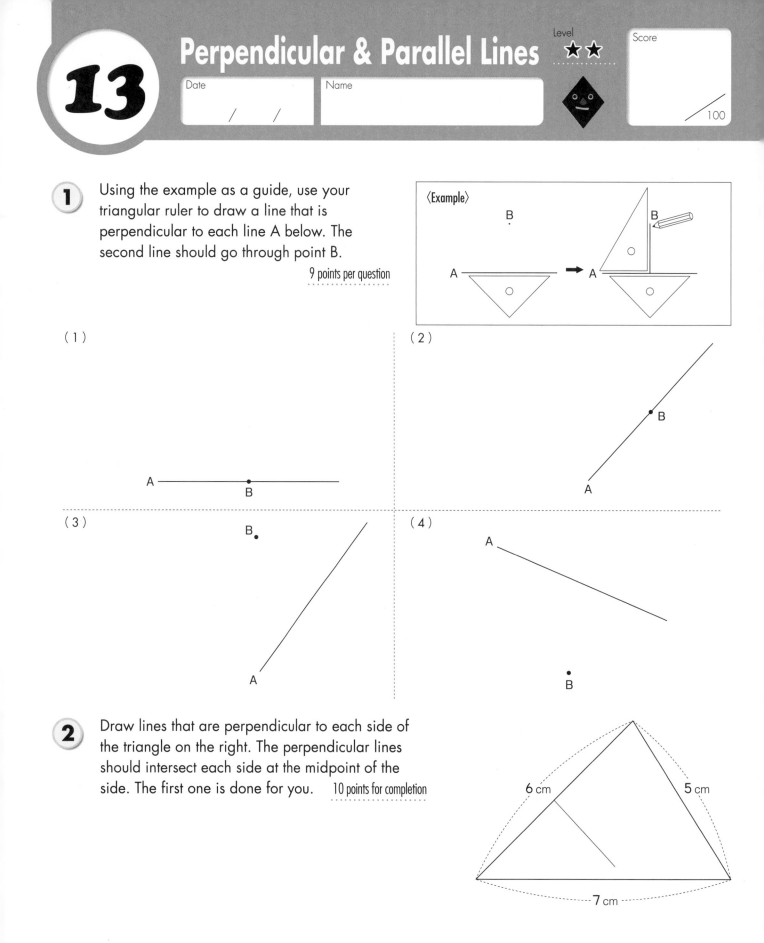

⟨Example⟩

(1)

(2)

(3)

(4)

2 Draw lines that are perpendicular to each side of the triangle on the right. The perpendicular lines should intersect each side at the midpoint of the side. The first one is done for you.   10 points for completion

6 cm   5 cm

7 cm

3 Using the example as a guide, use your triangular ruler to draw a line that is parallel to each line A below. The second line should go through point B.

9 points per question

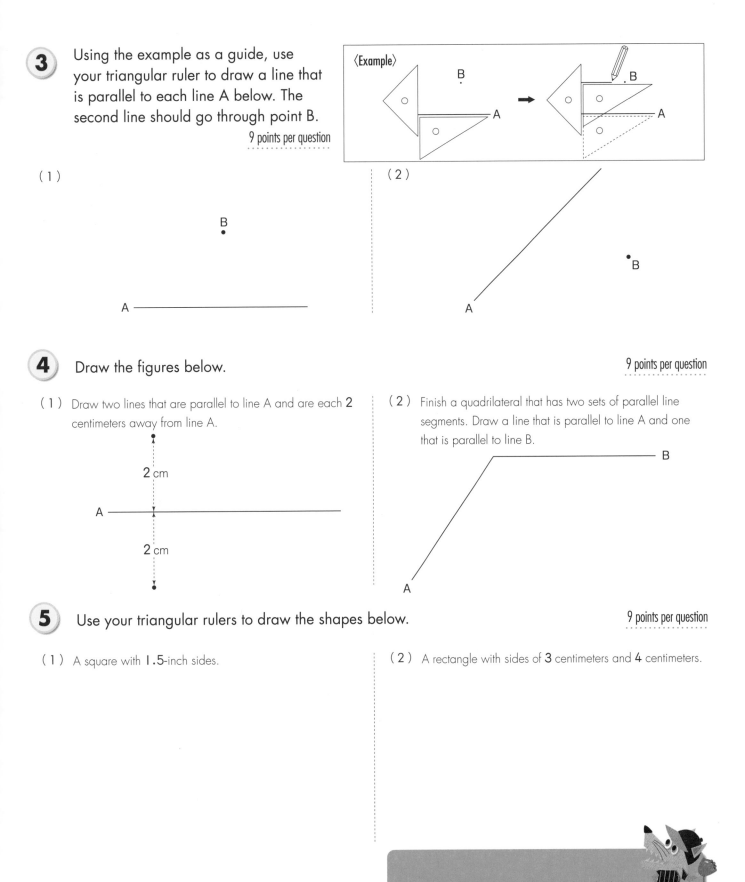

⟨Example⟩

(1)

B

A ——————————

(2)

A

B

4 Draw the figures below.

9 points per question

(1) Draw two lines that are parallel to line A and are each **2** centimeters away from line A.

2 cm

A ——————————

2 cm

(2) Finish a quadrilateral that has two sets of parallel line segments. Draw a line that is parallel to line A and one that is parallel to line B.

B

A

5 Use your triangular rulers to draw the shapes below.

9 points per question

(1) A square with 1.5-inch sides.

(2) A rectangle with sides of **3** centimeters and **4** centimeters.

Alright, let's try some quadrilaterals then!

Don't forget!

Trapezoids are quadrilaterals that have only one pair of parallel sides.

parallel

Trapezoid

---

1 Which lines are parallel in the shapes below?

10 points per question

(1)

A   D

B   C

(⎞⎟) and (⎞⎟)

(2)

A
   D
   C
B

(⎞⎟) and (⎞⎟)

(3)

A   D

B   C

(⎞⎟) and (⎞⎟)

(4)

A   D
B   C

(⎞⎟) and (⎞⎟)

---

2 Which of the shapes below are trapezoids?

15 points for completion

ⓐ   ⓑ   ⓒ

ⓓ   ⓔ   ⓕ

(⎞⎟)

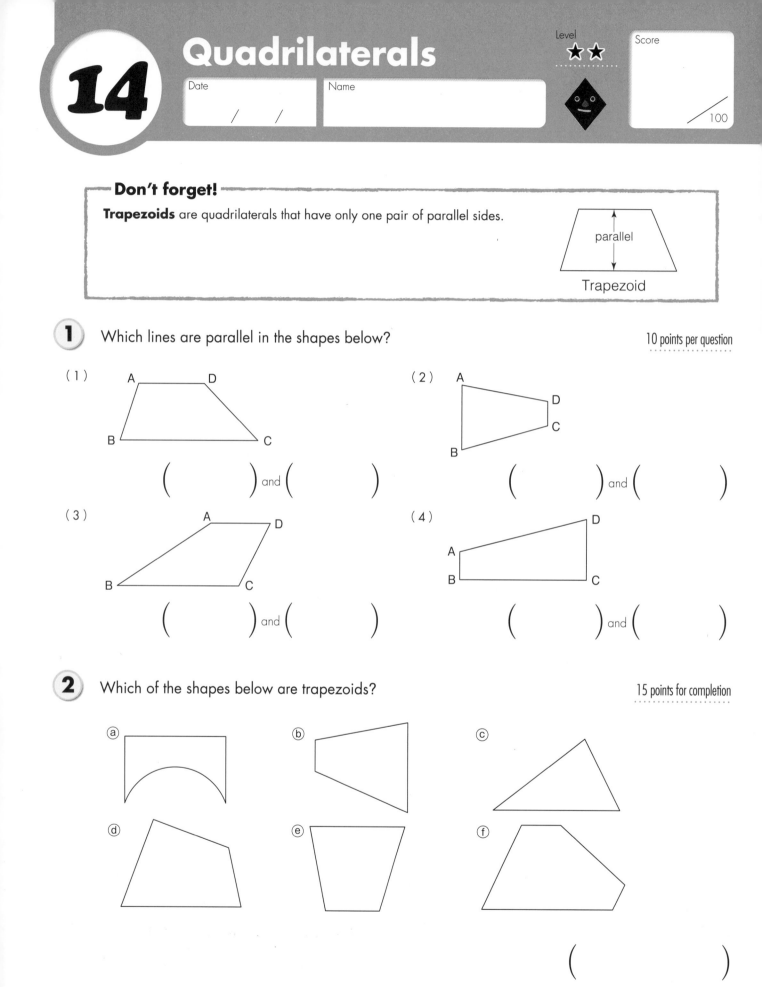

3 Draw the trapezoids below in the space provided. Use the example as a guide.

15 points per question

⟨Example⟩

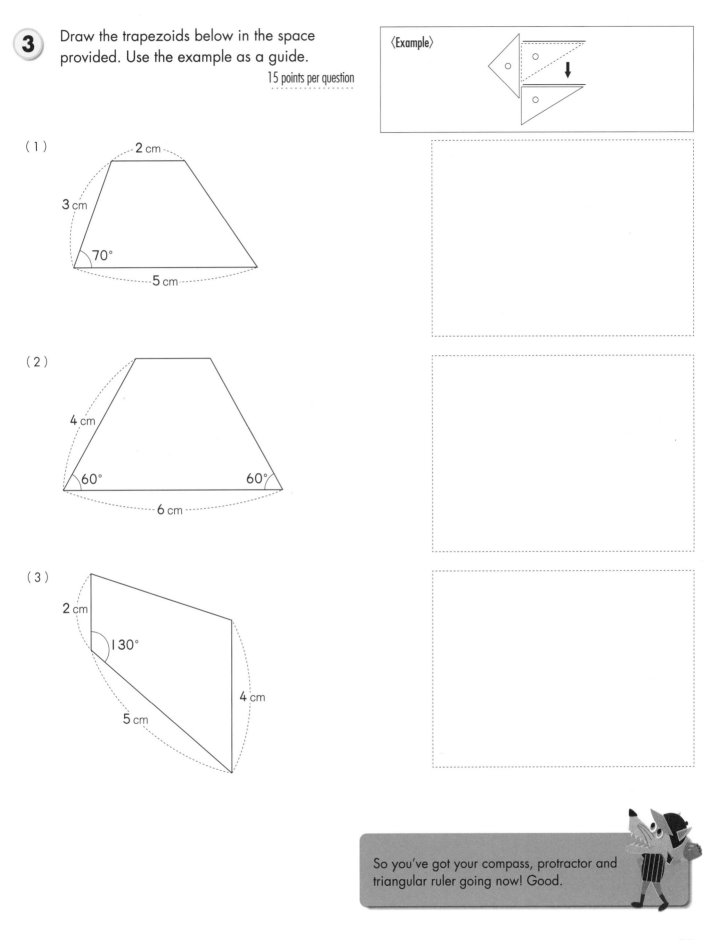

(1)

2 cm
3 cm
70°
5 cm

(2)

4 cm
60°   60°
6 cm

(3)

2 cm
130°
5 cm
4 cm

So you've got your compass, protractor and triangular ruler going now! Good.

Quadrilaterals

Date   /   /

Name

Level ★★

Score   /100

Don't forget!

Parallelograms are quadrilaterals with two sets of parallel sides.

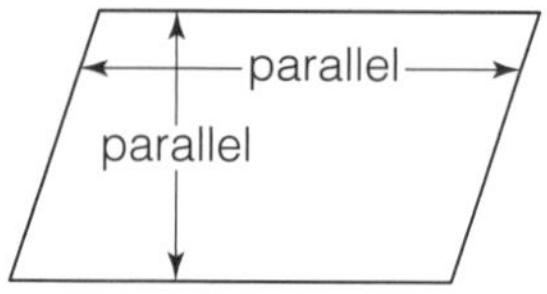

parallel

parallel

Parallelogram

1 The quadrilaterals below are parallelograms. Which sides are parallel?

6 points per question

(1)

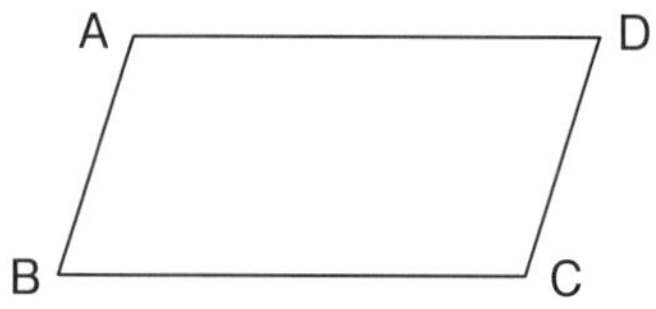

A   D
B   C

(   ) and (   )

(   ) and (   )

(2)

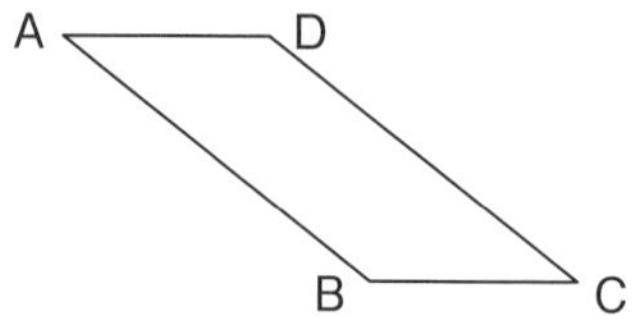

A   D
B   C

(   ) and (   )

(   ) and (   )

2 Which of the shapes below are parallelograms?

10 points for completion

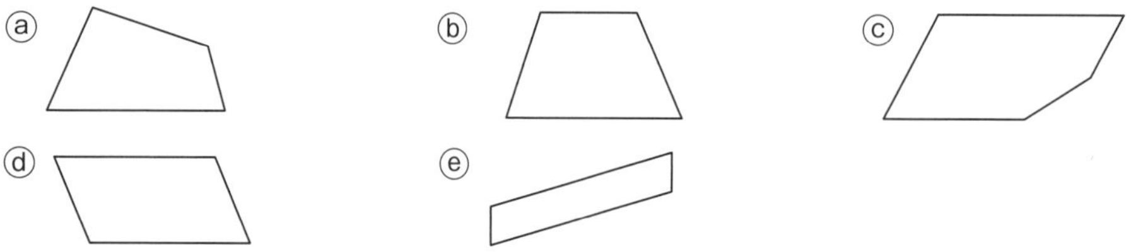

ⓐ   ⓑ   ⓒ

ⓓ   ⓔ

(   )

3 In parallelograms, the sum of the angles on each side of the shape is equal to 180°. This means that angles that are opposite to each other are equal. Use the figure on the right to answer the questions below.

6 points per question

(1) How long is side BC?   (   )

(2) How long is side CD?   (   )

(3) Find angle a.   (60°)

(4) Find angle b.   (   )

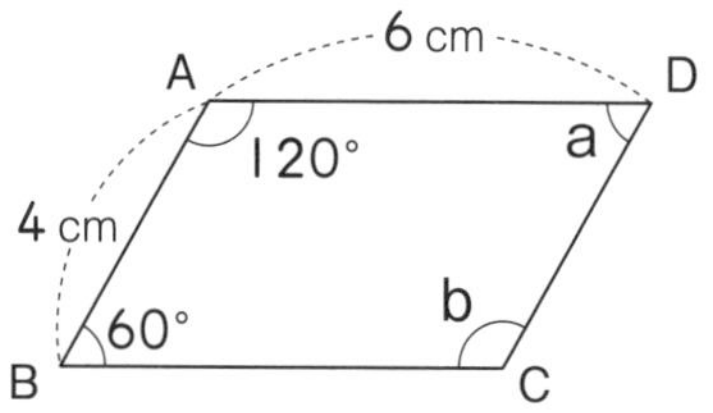

A ⸺ 6 cm ⸺ D
120°   a
4 cm
60°   b
B   C

Parallelogram

4 Answer the questions below about the parallelogram pictured on the right.

6 points per question

(1) How long is side AB?   (   )

(2) How long is side AD?   (   )

(3) Find angle a.   (   )

(4) Find angle b.   (   )

5 Draw each parallelogram below. Use your triangular rulers to draw the parallel lines as shown in the example on the right.

15 points per question

⟨Example⟩

(two parallel lines)

(1)

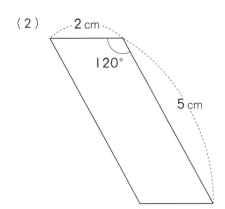

3 cm
70°
5 cm

(2)

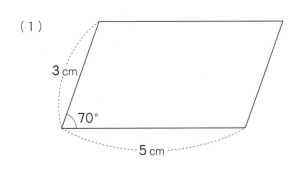

2 cm
120°
5 cm

This is tough. You're doing great!

16 Quadrilaterals

Date / /

Name

Don't forget!

A **rhombus** is a quadrilateral with four congruent sides.

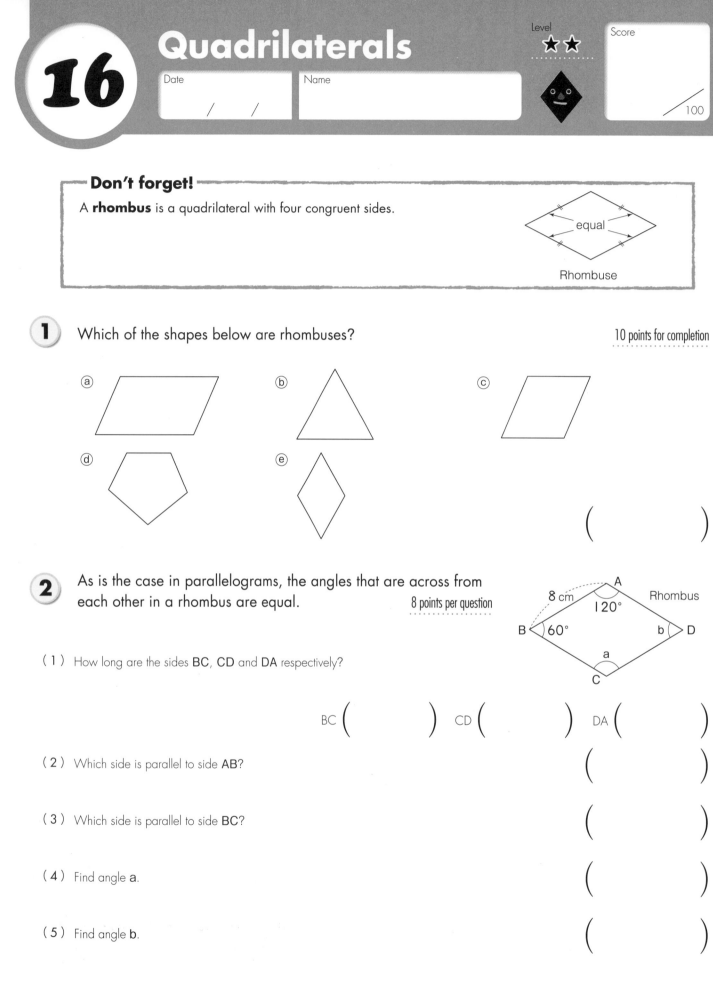

equal

Rhombuse

1 Which of the shapes below are rhombuses?

10 points for completion

ⓐ   ⓑ   ⓒ

ⓓ   ⓔ

( )

2 As is the case in parallelograms, the angles that are across from each other in a rhombus are equal.

8 points per question

8 cm   A   Rhombus
120°
B 60°   b D
a
C

(1) How long are the sides **BC**, **CD** and **DA** respectively?

BC ( ) CD ( ) DA ( )

(2) Which side is parallel to side **AB**?

( )

(3) Which side is parallel to side **BC**?

( )

(4) Find angle **a**.

( )

(5) Find angle **b**.

( )

32   © Kumon Publishing Co., Ltd.

3 Draw the rhombuses below.

15 points per question

(1)

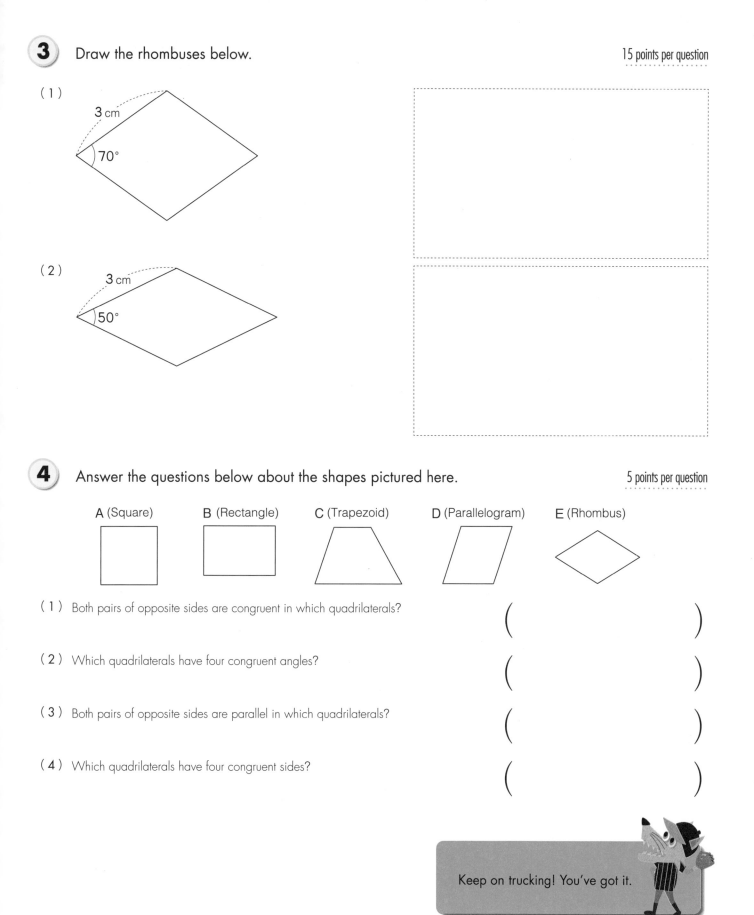

3 cm
70°

(2)

3 cm
50°

4 Answer the questions below about the shapes pictured here.

5 points per question

A (Square)   B (Rectangle)   C (Trapezoid)   D (Parallelogram)   E (Rhombus)

(1) Both pairs of opposite sides are congruent in which quadrilaterals?

(   )

(2) Which quadrilaterals have four congruent angles?

(   )

(3) Both pairs of opposite sides are parallel in which quadrilaterals?

(   )

(4) Which quadrilaterals have four congruent sides?

(   )

Keep on trucking! You've got it.

Quadrilaterals

1 Connect both sets of opposite vertices as shown in the example.

5 points per question

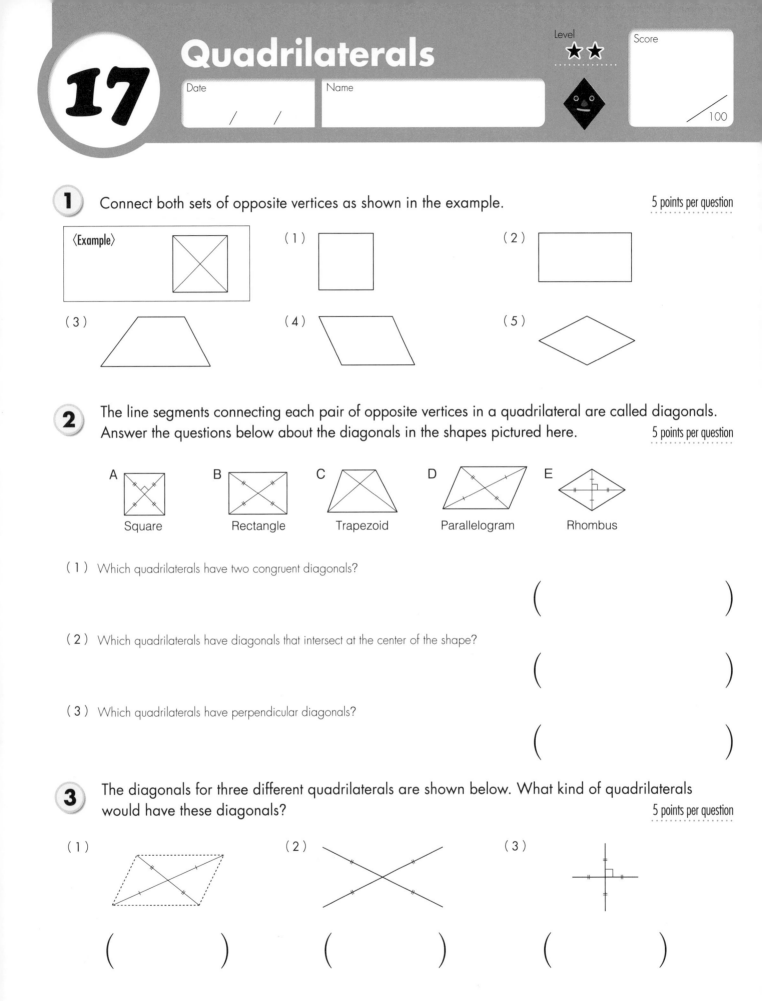

〈Example〉

(1)

(2)

(3)

(4)

(5)

2 The line segments connecting each pair of opposite vertices in a quadrilateral are called diagonals. Answer the questions below about the diagonals in the shapes pictured here.

5 points per question

A Square B Rectangle C Trapezoid D Parallelogram E Rhombus

(1) Which quadrilaterals have two congruent diagonals?

(   )

(2) Which quadrilaterals have diagonals that intersect at the center of the shape?

(   )

(3) Which quadrilaterals have perpendicular diagonals?

(   )

3 The diagonals for three different quadrilaterals are shown below. What kind of quadrilaterals would have these diagonals?

5 points per question

(1)

(2)

(3)

(   )   (   )   (   )

4 Answer the questions below about the parallelogram pictured on the right.

5 points per question

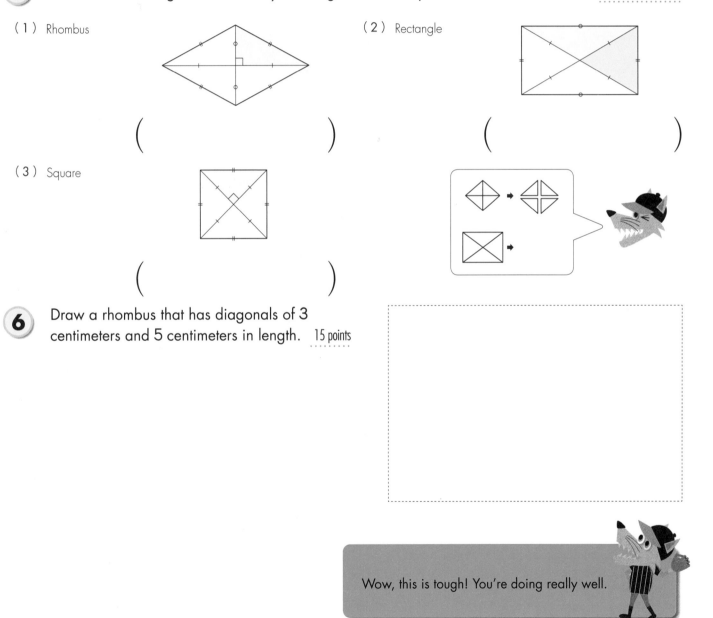

(1) How long is the line segment ED?

(   )

(2) How long is the diagonal BD?

(   )

(3) How long is the line EC?

(   )

5 What kind of triangles are made by the diagonals in the quadrilaterals below?

5 points per question

(1) Rhombus

(   )

(2) Rectangle

(   )

(3) Square

(   )

6 Draw a rhombus that has diagonals of 3 centimeters and 5 centimeters in length. 15 points

Wow, this is tough! You're doing really well.

18

Angles

Level ★★

Score

/100

Date   /   /

Name

Don't forget!

The sum of internal angles in a triangle is 180°.

Angle A + Angle B + Angle C = 180°

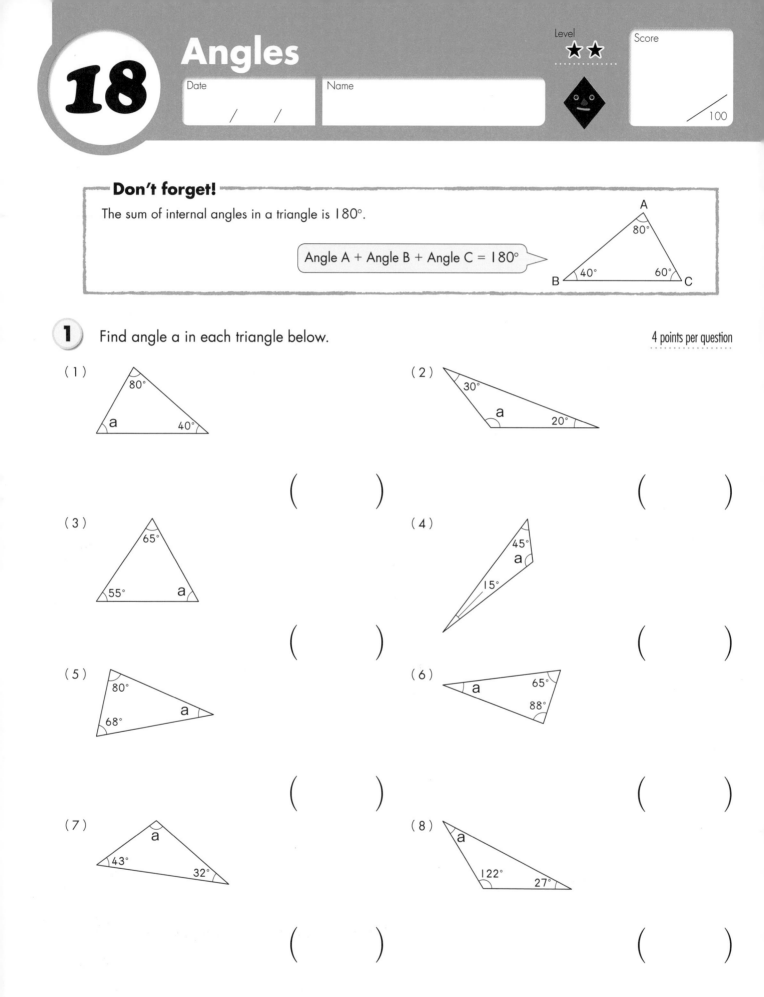

1   Find angle a in each triangle below.

4 points per question

(1)

(   )

(2)

(   )

(3)

(   )

(4)

(   )

(5)

(   )

(6)

(   )

(7)

(   )

(8)

(   )

2 Find angle a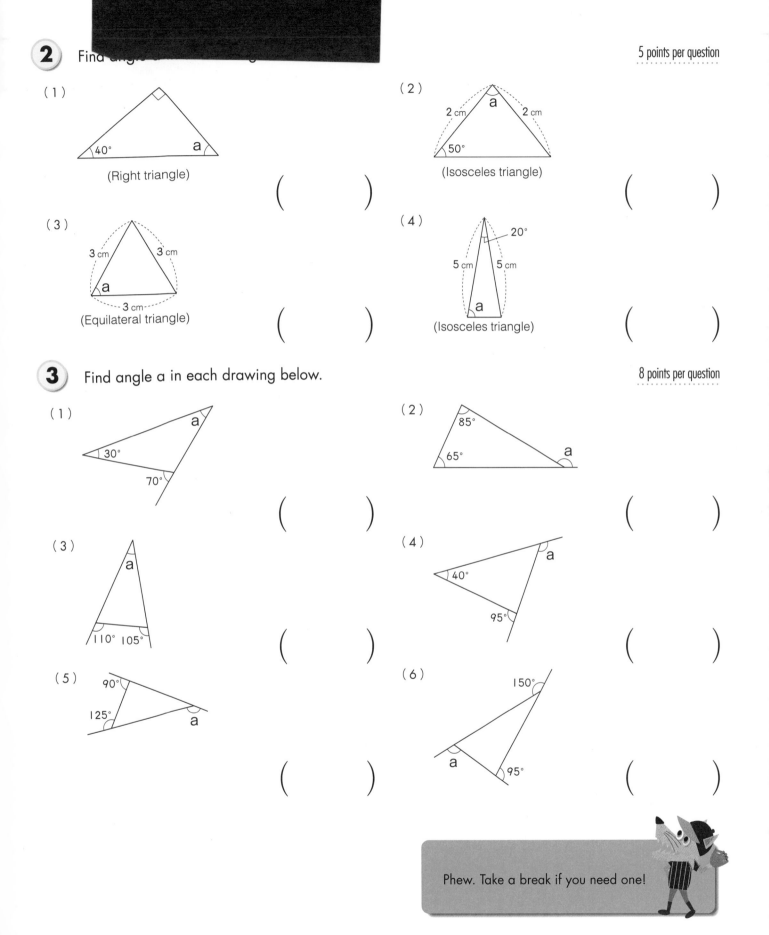

（1）

(Right triangle)

40°   a

(   )

（2）

2 cm   a   2 cm

50°

(Isosceles triangle)

(   )

（3）

3 cm   3 cm

a

3 cm

(Equilateral triangle)

(   )

（4）

20°

5 cm   5 cm

a

(Isosceles triangle)

(   )

3 Find angle a in each drawing below.

（1）

a

30°

70°

(   )

（2）

85°

65°   a

(   )

（3）

a

110° 105°

(   )

（4）

a

40°

95°

(   )

（5）

90°

125°   a

(   )

（6）

150°

a   95°

(   )

Phew. Take a break if you need one!

Date   /   /

Name

/100

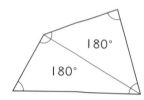

1 A quadrilateral can be separated into two triangles by a diagonal. What is the sum of internal angles of a quadrilateral?

6 points

(   )

2 Find angle a in each quadrilateral below.

6 points per question

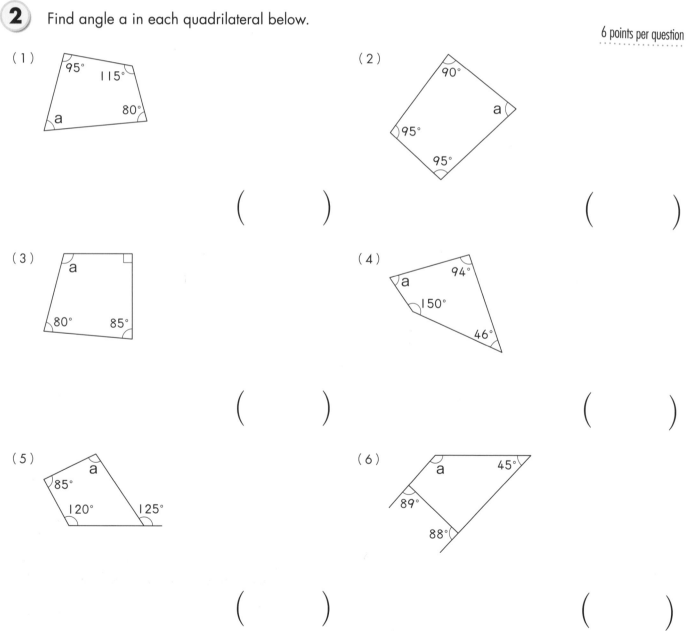

(1) 95° 115° a 80°

(   )

(2) 90° a 95° 95°

(   )

(3) a 80° 85°

(   )

(4) a 94° 150° 46°

(   )

(5) a 85° 120° 125°

(   )

(6) a 45° 89° 88°

(   )

3 A pentagon can be separated into three triangles by diagonals. What is the sum of the internal angles in a pentagon?

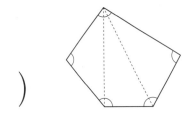

8 points

( )

4 Find angle a in each figure below.

10 points per question

(1)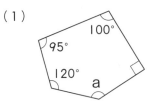

100°
95°
120°
a

( )

(2)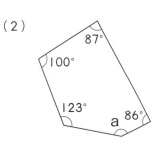

87°
100°
123°
a 86°

( )

 5 What is the sum of internal angles in a hexagon?

10 points

( )

6 Find angle a in each figure below.

10 points per question

(1)
100°
155°   95°
a
145°   110°

( )

(2)
a   85°
150°
145°
134°
116°

( )

Okay, time to add more sides to your shapes.
Are you ready?

Don't forget!

Any shape made up of many line segments connecting around a space is called a **polygon**. A polygon in which all the sides are congruent and all the angles are congruent is called a **regular polygon**. The polygons on the right are regular polygons.

Equilateral triangle   Square   Regular pentagon

1 Write the name of each regular polygon below.

4 points per question

(1)   (2)   (3)   (4)

(   )   (   )   (   )   (   )

2 Write a ✓ under regular polygons, and an ✕ under irregular polygons below.

3 points per question

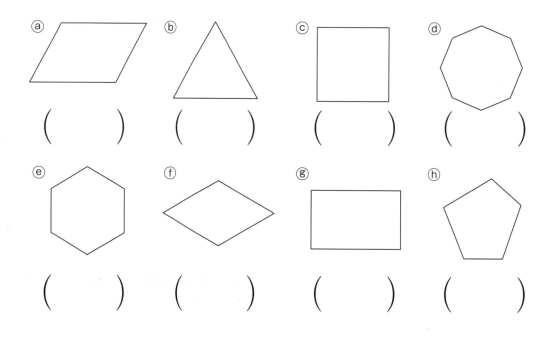

ⓐ   ⓑ   ⓒ   ⓓ

(   )   (   )   (   )   (   )

ⓔ   ⓕ   ⓖ   ⓗ

(   )   (   )   (   )   (   )

3 Each polygon below is regular and was drawn by dividing up a circle into equal parts. Find angle a for each polygon.

10 points per question

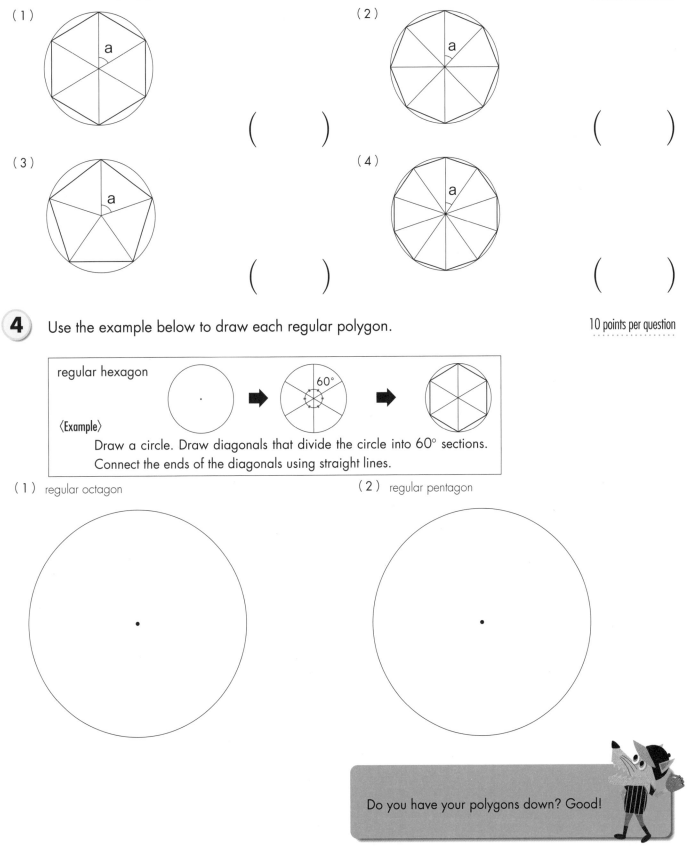

(1)

( )

(2)

( )

(3)

( )

(4)

( )

4 Use the example below to draw each regular polygon.

10 points per question

regular hexagon

60°

〈Example〉
Draw a circle. Draw diagonals that divide the circle into 60° sections. Connect the ends of the diagonals using straight lines.

(1) regular octagon

(2) regular pentagon

Do you have your polygons down? Good!

21

Congruent Figures

Level ★★

Date / /

Name

Score

/100

1 Which figures are the same size and shape as triangle A?

10 points for completion

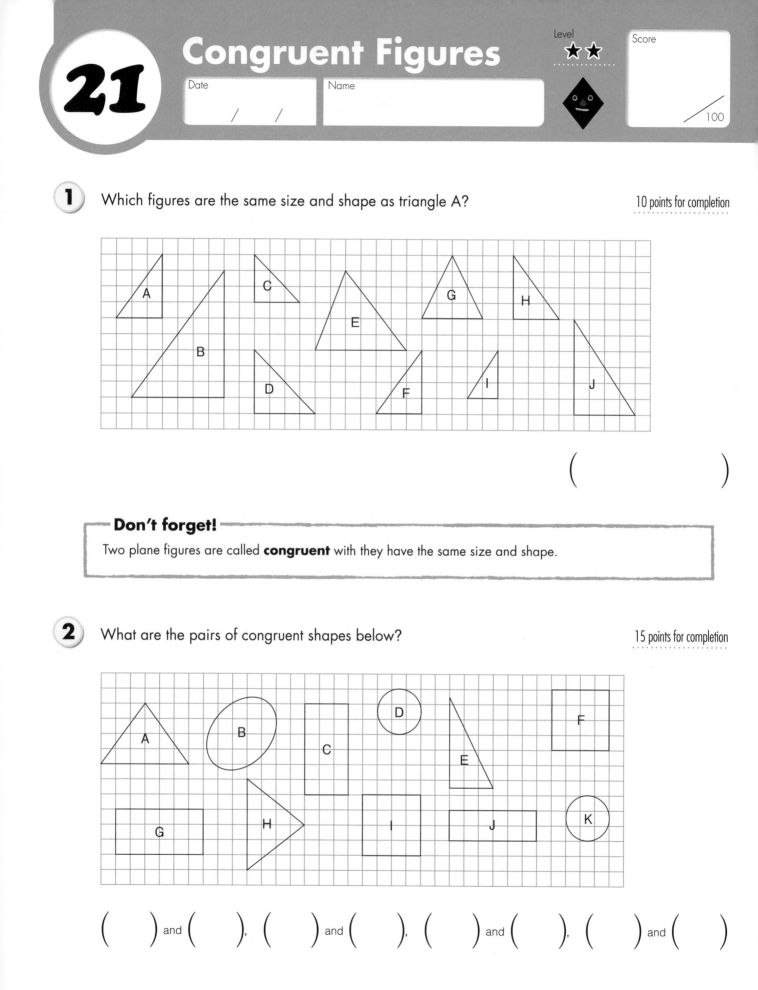

( )

Don't forget!

Two plane figures are called **congruent** with they have the same size and shape.

2 What are the pairs of congruent shapes below?

15 points for completion

( ) and ( ), ( ) and ( ), ( ) and ( ), ( ) and ( )

3 In congruent figures, the vertices, sides and angles that are the same in each figure are called corresponding vertices, sides and angles. Use the figures on the right to answer the questions below.

5 points per question

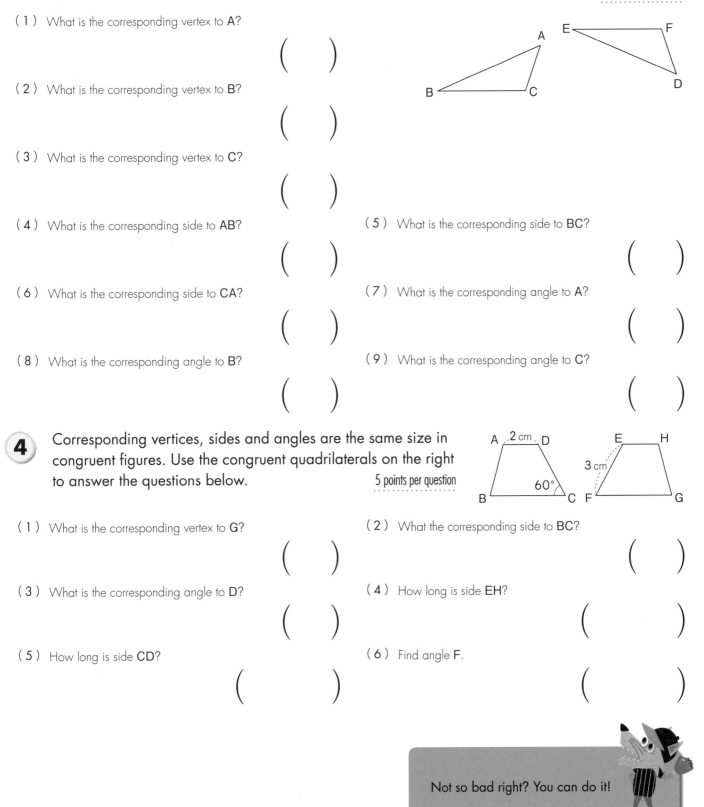

(1) What is the corresponding vertex to A?

(   )

(2) What is the corresponding vertex to B?

(   )

(3) What is the corresponding vertex to C?

(   )

(4) What is the corresponding side to AB?

(   )

(5) What is the corresponding side to BC?

(   )

(6) What is the corresponding side to CA?

(   )

(7) What is the corresponding angle to A?

(   )

(8) What is the corresponding angle to B?

(   )

(9) What is the corresponding angle to C?

(   )

4 Corresponding vertices, sides and angles are the same size in congruent figures. Use the congruent quadrilaterals on the right to answer the questions below.

5 points per question

(1) What is the corresponding vertex to G?

(   )

(2) What the corresponding side to BC?

(   )

(3) What is the corresponding angle to D?

(   )

(4) How long is side EH?

(   )

(5) How long is side CD?

(   )

(6) Find angle F.

(   )

Not so bad right? You can do it!

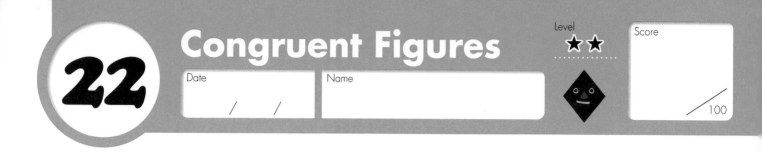

Congruent Figures

22

Date ____ / ____ / ____　　Name _____

1　Compare the two triangles (triangle ABC and triangle CDA) made by dividing up the parallelogram ABCD.

5 points per question

(1)　Which side of the parallelogram is congruent to **AB**?

(　　　　　)

(2)　Which side of the parallelogram is congruent to **BC**?

(　　　　　)

(3)　Which line is congruent to **CA**?

(　　　　　)

(4)　Which angle is congruent to **a**?

(　　　　　)

(5)　Which angle is congruent to **b**?

(　　　　　)

(6)　Which angle is congruent to **c**?

(　　　　　)

(7)　Are the two triangles congruent?

(　　　　　)

In the parallelogram ABCD, AD and BC are parallel, so angle *a* = angle *c* and angle *b* = angle *d*.

2　When a parallelogram is divided by a diagonal, two congruent triangles are made. What are the two congruent triangles in this illustration?

11 points

(　　　　　) and (　　　　　)

3　When a parallelogram is divided by two diagonals, four triangles are made. Use the figure to the right to answer the questions below.

6 points per question

(1)　Which triangle is congruent to **ABO**?

(　　　　　)

(2)　Which triangle is congruent to triangle **BCO**?

(　　　　　)

4 In the figure on the right, a rectangle has been divided by two diagonals. Use it to answer the questions below.

6 points per question

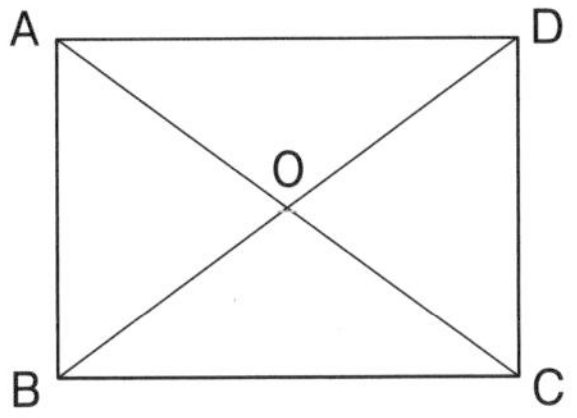

(1) Which triangles are congruent to ABD?

( )

(2) Which triangle is congruent to ABO?

( )

(3) Which triangle is congruent to BCO?

( )

5 In the figure on the right, a rhombus has been divided by two diagonals. Use it to answer the questions below.

6 points per question

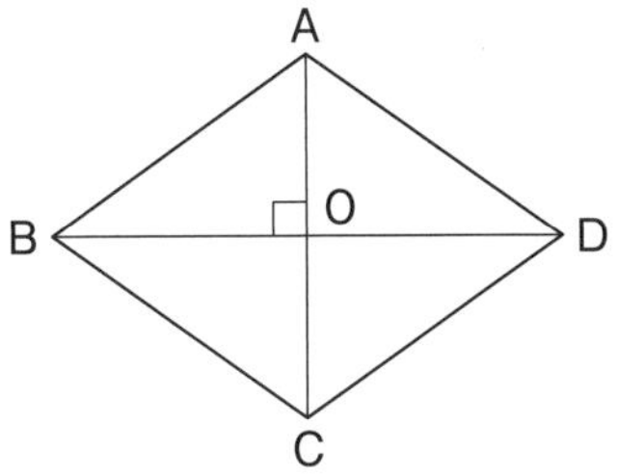

(1) Which triangle is congruent to ABD?

( )

(2) Which triangles are congruent to ABO?

( ) and ( ) and ( )

6 In the figure on the right, a square has been divided by two diagonals. Use it to answer the questions below.

6 points per question

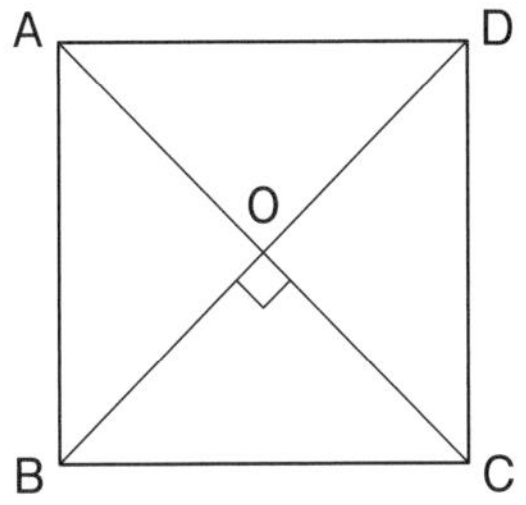

(1) Which triangles are congruent to ABO?

( ) and ( ) and ( )

(2) Which triangles are congruent to ABD?

( ) and ( ) and ( )

Good job! Let's switch it up now.

23 Axial Symmetry

Date   /   /   Name

1 If you fold the following shapes in half, which of them will overlap perfectly? Write a ✓ below the correct shapes.

10 points for completion

A   (   )   B   (   )   C   (   )   D   (   )

> **Don't forget!**
>
> If you can find a way to fold a shape in half so that both halves overlap perfectly, that shape has **axial symmetry**. That means that the form of the shape is the same on both sides of one axis.

2 Which shapes below have axial symmetry? Write a ✓ below the correct shapes.

10 points for completion

A   (   )   B   (   )   C   (   )   D   (   )

3 Which shapes below have axial symmetry? Write a ✓ below the correct shapes.

20 points for completion

A Rectangle   (   )   B Parallelogram   (   )   C Rhombus   (   )   D Square   (   )

E Isosceles triangle   (   )   F Equilateral triangle   (   )   G Right triangle   (   )   H Equilateral right triangle   (   )

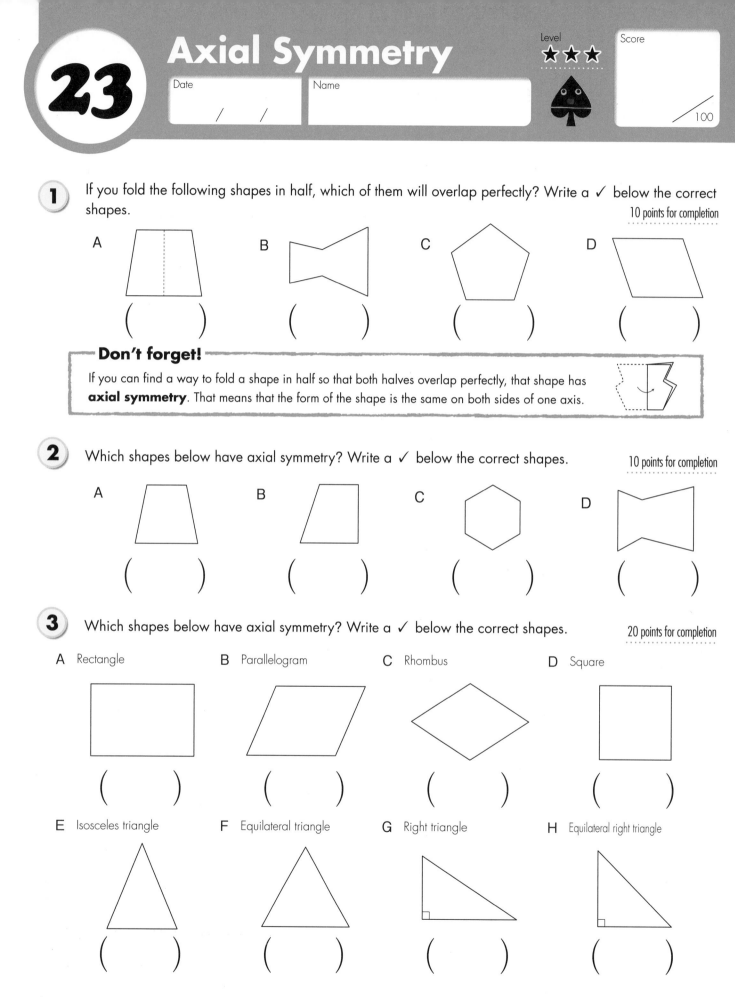

4 Each of the shapes below has axial symmetry. Where could you fold each shape in half exactly?
Draw the folding line on each shape.

5 points per question

(1)　　　　　　　　(2)　　　　　　　　(3)　　　　　　　　(4)

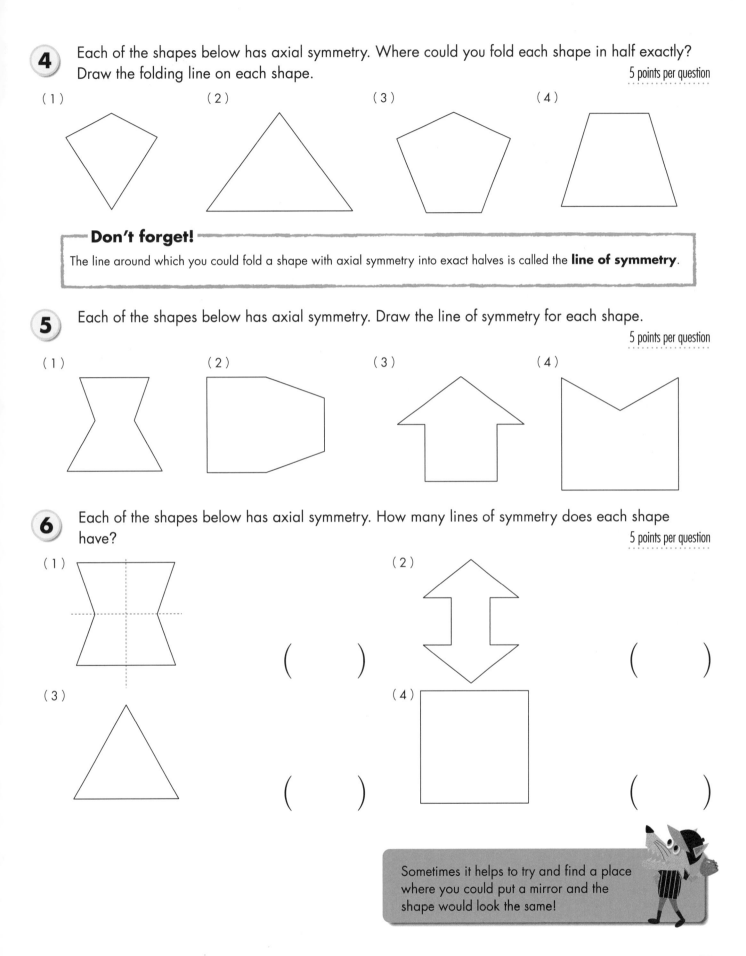

Don't forget!

The line around which you could fold a shape with axial symmetry into exact halves is called the **line of symmetry**.

5 Each of the shapes below has axial symmetry. Draw the line of symmetry for each shape.

5 points per question

(1)　　　　　　　　(2)　　　　　　　　(3)　　　　　　　　(4)

6 Each of the shapes below has axial symmetry. How many lines of symmetry does each shape
have?

5 points per question

(1)　　　　　　　　　　　　　　　　(2)

(　　　)　　　　　　　　　　　　　　　　(　　　)

(3)　　　　　　　　　　　　　　　　(4)

(　　　)　　　　　　　　　　　　　　　　(　　　)

Sometimes it helps to try and find a place where you could put a mirror and the shape would look the same!

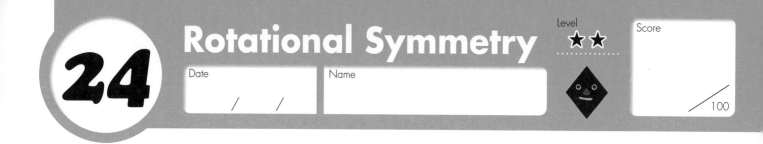
1 If you rotate each shape below 180°, which of them would overlap the original shape exactly? Write a ✓ next to the correct shapes.

20 points for completion

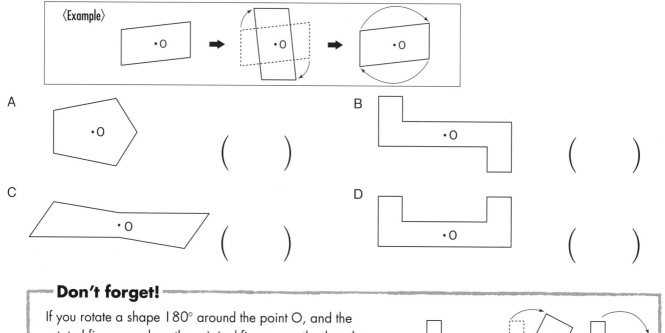

A ( )

B ( )

C ( )

D ( )

Don't forget!

If you rotate a shape 180° around the point O, and the rotated figure overlaps the original figure exactly, then that shape has **rotational symmetry**.

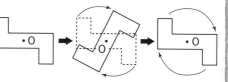

---

2 Write a ✓ next to each shape below that has rotational symmetry.

20 points for completion

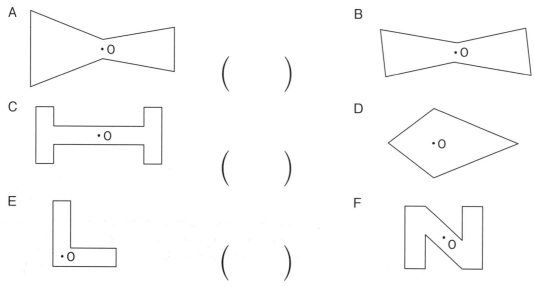

A ( )

B ( )

C ( )

D ( )

E ( )

F ( )

3 When a figure with rotational symmetry is spun 180° round the center of symmetry, the overlapping vertices, sides and angles are called corresponding vertices, sides and angles. Answer the questions below about the rotationally symmetric figure pictured here.

5 points per question

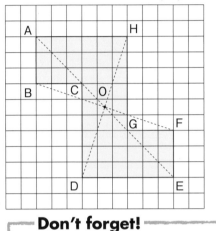

(1) What is the corresponding vertex to A? ( )

(2) What is the corresponding vertex to B? ( )

(3) What is the corresponding side to AH? ( )

(4) What is the corresponding angle to D? ( )

(5) The lines **AE**, **BF**, **CG** and **DH** all pass through what point? ( )

(6) What line is the same size as OA? ( )

(7) What line is the same size as OB? ( )

(8) What line is the same size as OD? ( )

4 The figure pictured here has rotational symmetry around the center of symmetry O. Answer the questions below.

4 points per question

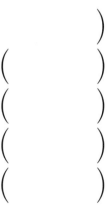

(1) What is the name of the point that lines **BG** and **CH** both cross through? ( )

(2) What is the length of HI? ( )

(3) If angle A is 110°, what is angle F? ( )

(4) If OB is **3.2** centimeters long, how long is OG? ( )

(5) If CH is **5.4** centimeters long, how long is OH? ( )

This is tough. Let's keep trying!

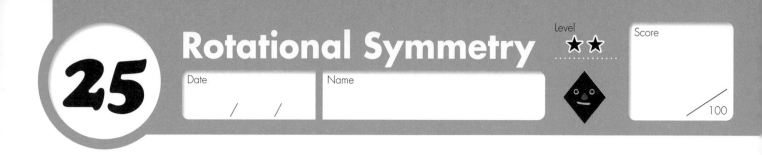

Rotational Symmetry

Date   /   /

Name

Level ★★

Score   /100

Don't forget!

In a figure with rotational symmetry, lines connecting all the corresponding vertices will all cross through the center of symmetry.

In the figure on the left, the lines AD, BE and CF all cross through the center of symmetry, O.

---

1 The figures below have rotational symmetry. Connect the corresponding vertices in order to find the center of symmetry, O.

10 points per question

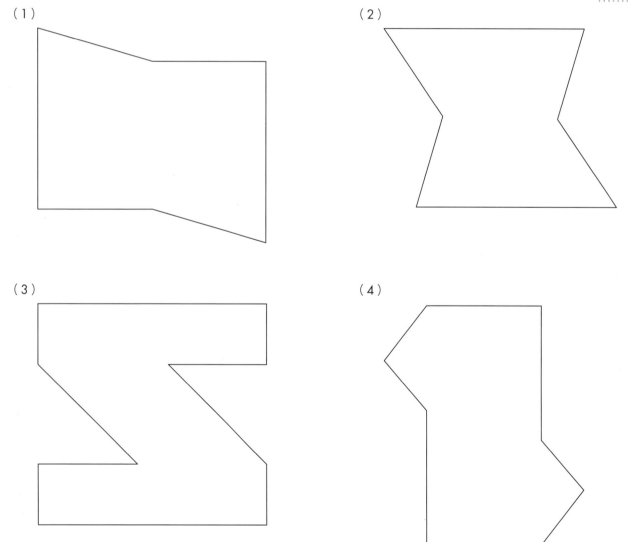

(1)

(2)

(3)

(4)

© Kumon Publishing Co., Ltd.

2 Each of the figures below represent half of a figure with rotational symmetry and a center of symmetry of O. First, draw each line of that passes through a vertex and the center of symmetry. Second, find each corresponding vertex. Third, connect all of the vertices in order to complete the figure.

15 points per question

(1)

(2)

(3)

(4)

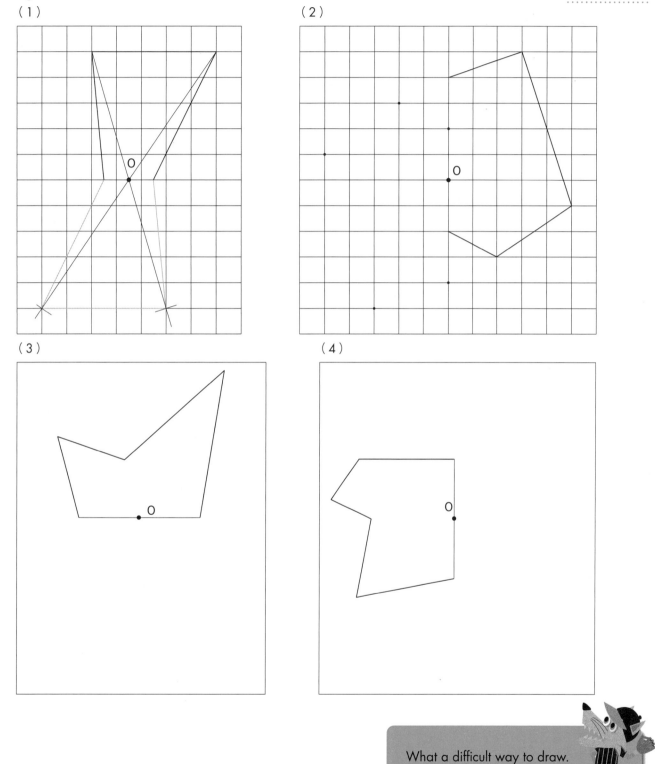

What a difficult way to draw. You're doing great!

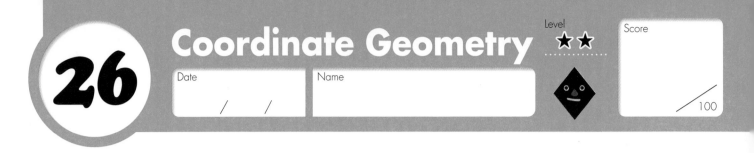

Date / / 　Name

Score
/100

Don't forget!

Point A on the graph to the right can be represented as (2, 3).
This means that the point is 2 over on the horizontal axis and 3
up on the vertical axis.

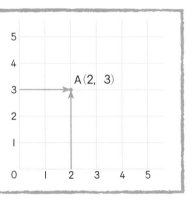

1 What are the coordinates of each point on the graph below?

5 points per point

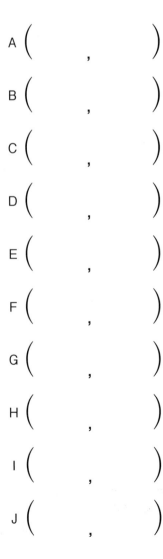

A (,)

B (,)

C (,)

D (,)

E (,)

F (,)

G (,)

H (,)

I (,)

J (,)

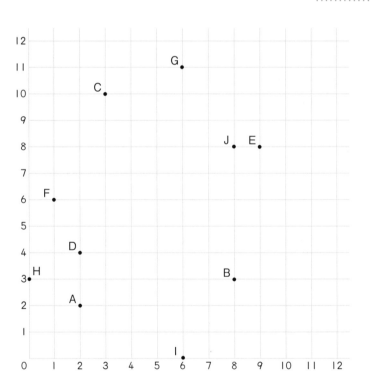

2 Mark each point on the graph below.

5 points per point

A (5, 1)   B (1, 5)   C (2, 7)   D (3, 5)   E (0, 10)

F (2, 0)   G (8, 2)   H (6, 6)   I (11, 3)   J (4, 9)

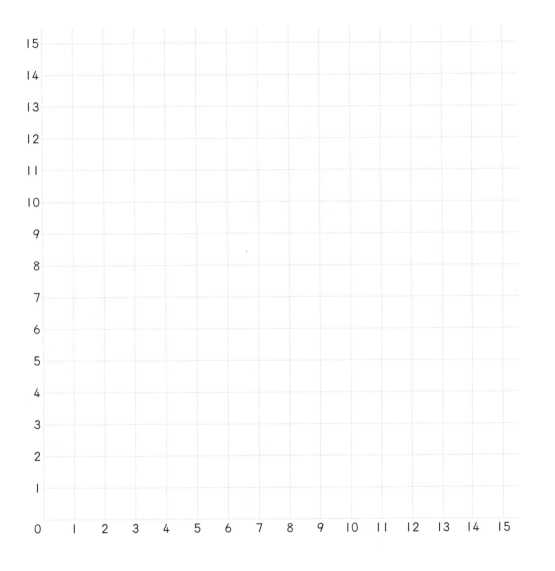

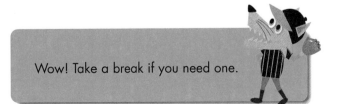

Wow! Take a break if you need one.

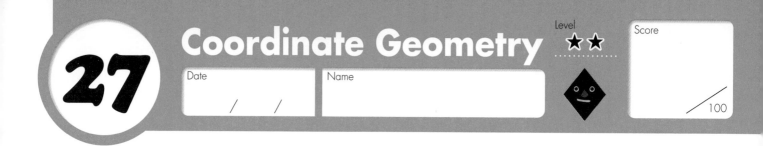

Coordinate Geometry

Date / /

Name

Level ★★

Score /100

1 Draw the shapes made by the following coordinate points on the graph below.

15 points per question

(1) A (2, 1), B (6, 1), C (6, 5)

(2) D (10, 6), E (7, 2), F (12, 2)

(3) G (0, 13), H (0, 8), I (5, 8), J (5, 13)

(4) K (5, 14), L (9, 7), M (14, 7), N (10, 14)

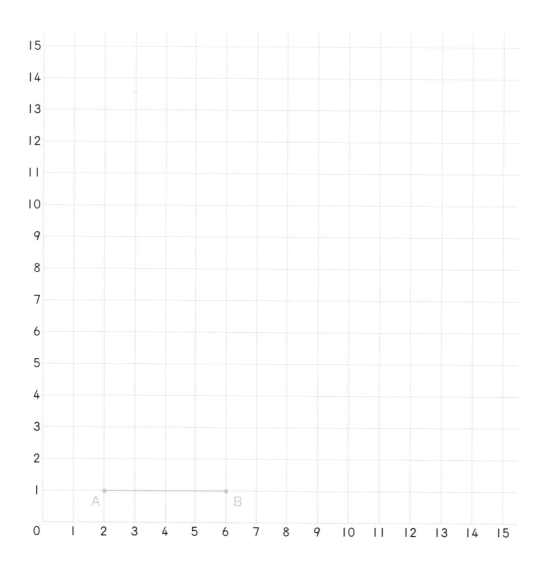

2 Draw new figures so that each graph has axial symmetry around the brown line.

(1)

(2)

(3)

(4)

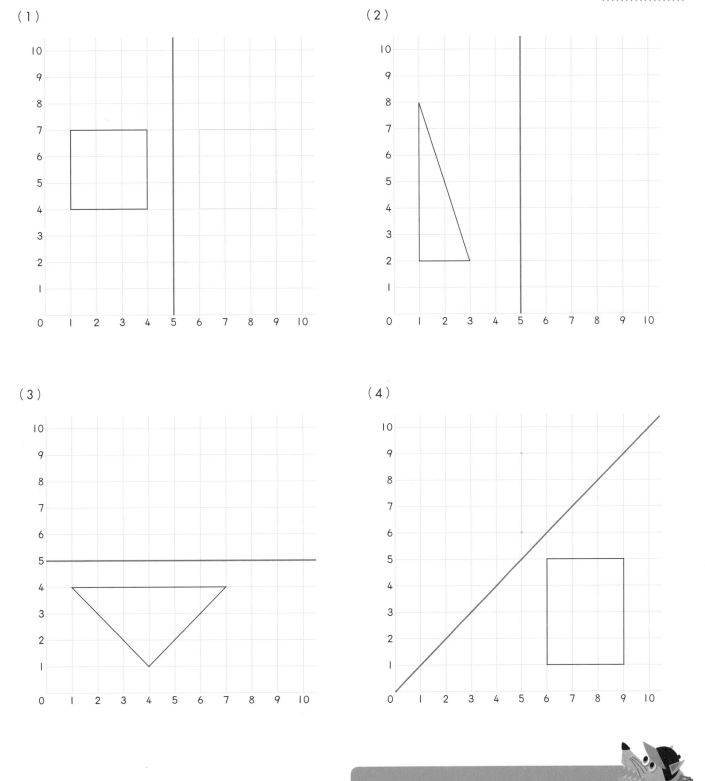

Remember the mirror! Each side should look like a reflection over the blue line.

28 Area

Date / /

Name

Don't forget!

The area of a parallelogram is the base times the height.

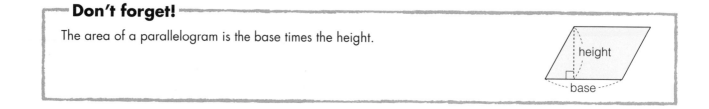

height

base

---

1 Find the area of each parallelogram below.

6 points per question

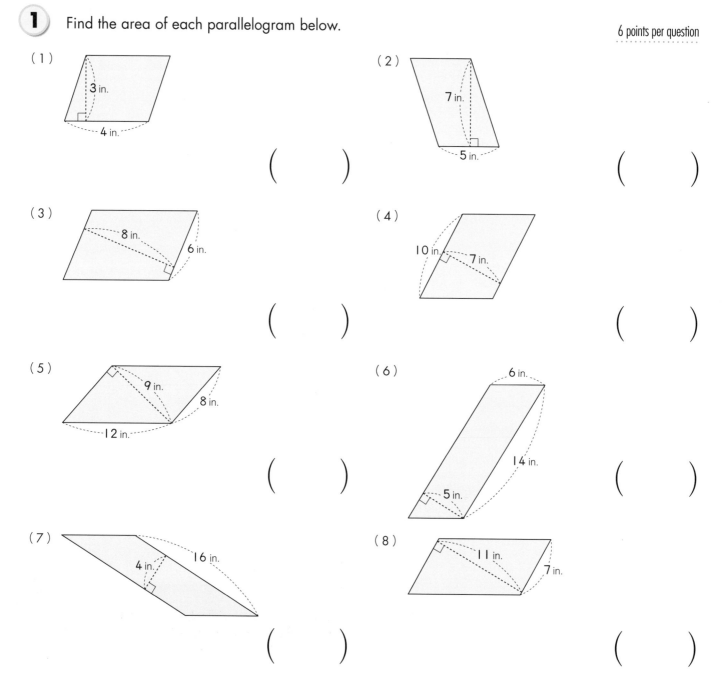

(1)

3 in.

4 in.

(   )

(2)

7 in.

5 in.

(   )

(3)

8 in.

6 in.

(   )

(4)

10 in.

7 in.

(   )

(5)

9 in.

8 in.

12 in.

(   )

(6)

6 in.

14 in.

5 in.

(   )

(7)

4 in.

16 in.

(   )

(8)

11 in.

7 in.

(   )

2 Answer the questions below about the parallelogram pictured here.

8 points per question

(1) What is the height of the parallelogram if you think of **AB** as the base?

(   )

(2) Find the area of this parallelogram using **AB** as the base.

(   )

3 Find the area of each parallelogram below.

6 points per question

(1)

(   )

(2)

(   )

(3)

(   )

(4)

(   )

(5)

(   )

(6)

(   )

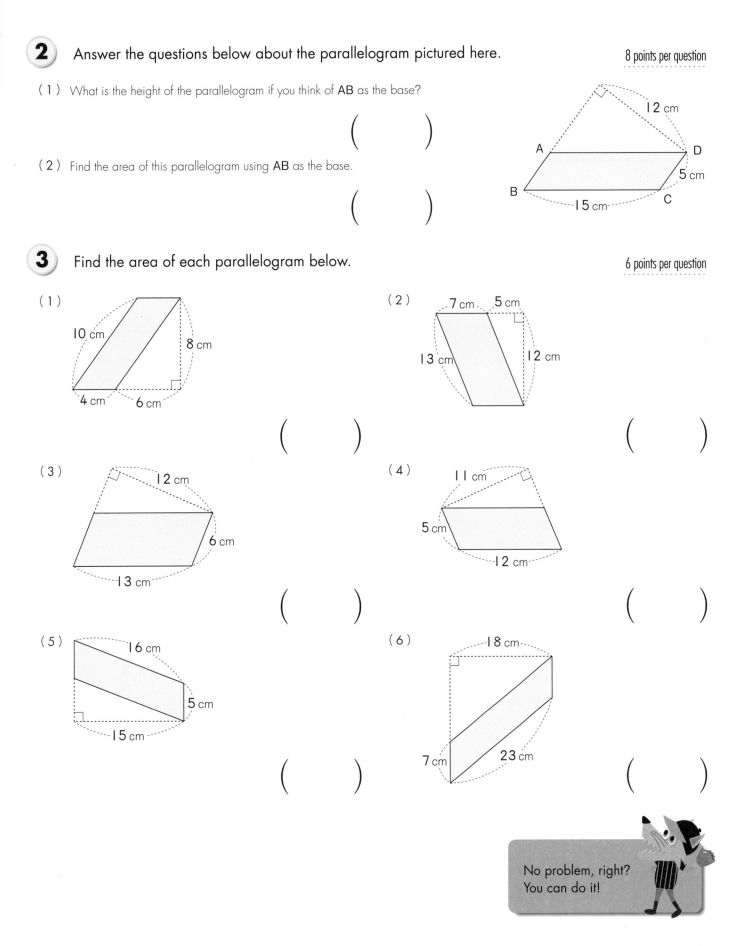

No problem, right?
You can do it!

Area

Don't forget!

The area of a triangle is the base times the height divided by 2.

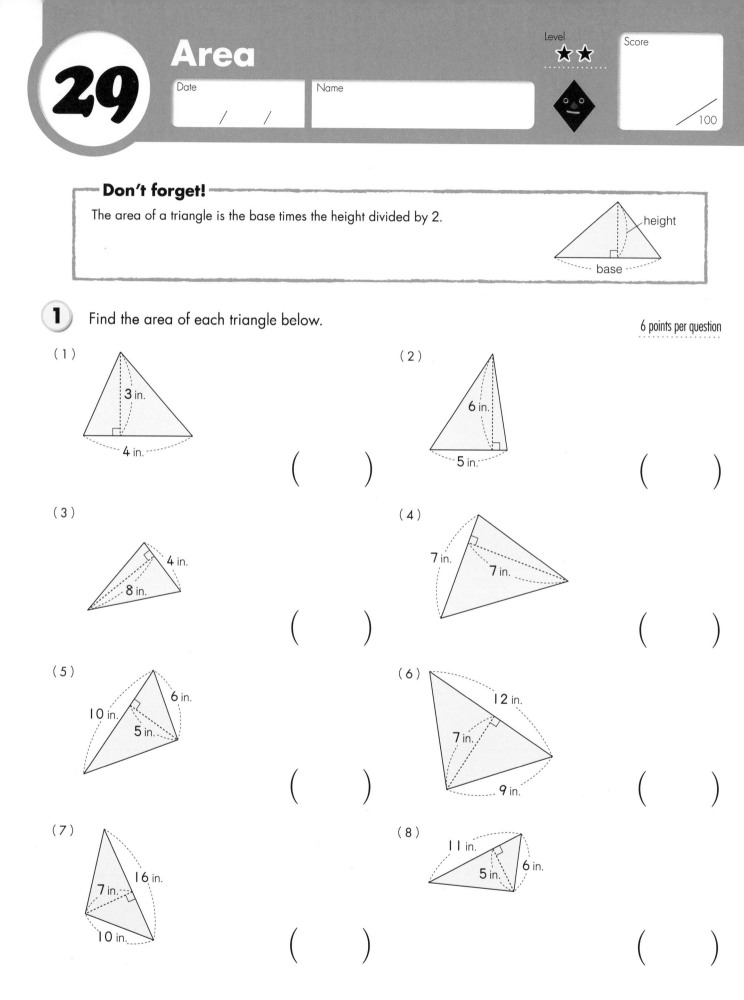

1 Find the area of each triangle below.

6 points per question

(1)
3 in.
4 in.

(   )

(2)
6 in.
5 in.

(   )

(3)
4 in.
8 in.

(   )

(4)
7 in.
7 in.

(   )

(5)
10 in.
6 in.
5 in.

(   )

(6)
12 in.
7 in.
9 in.

(   )

(7)
16 in.
7 in.
10 in.

(   )

(8)
11 in.
5 in.
6 in.

(   )

© Kumon Publishing Co., Ltd.

2 Answer the questions below about the triangle pictured here.

8 points per question

(1) What is the height of the triangle if you think of BC as the base?

(   )

(2) Find the area of the triangle using BC as the base.

(   )

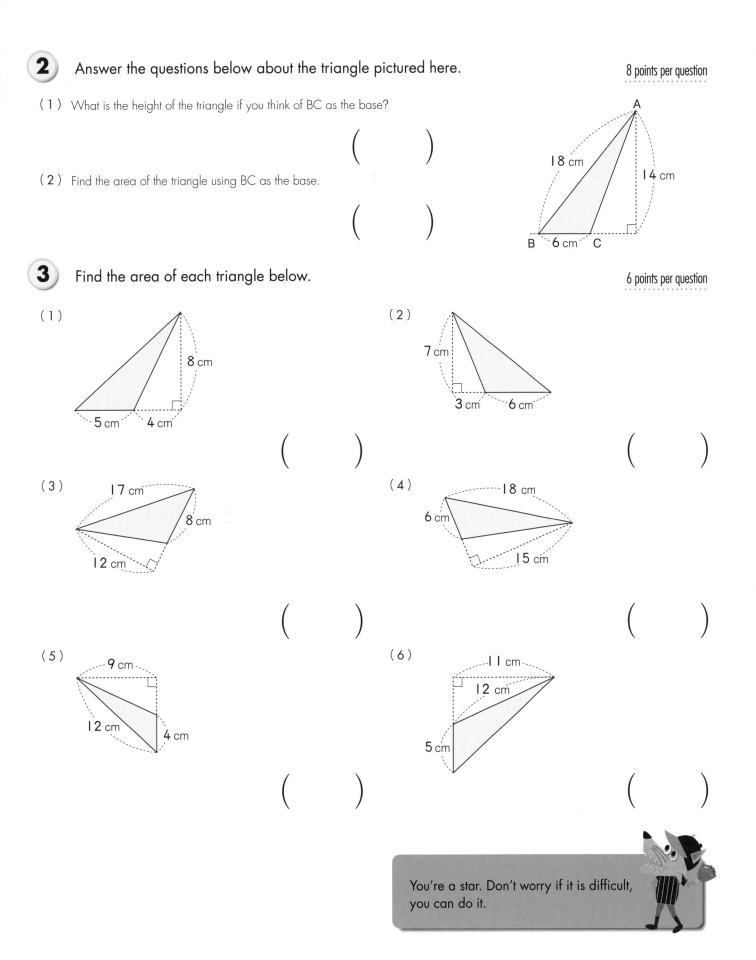

3 Find the area of each triangle below.

6 points per question

(1)

(   )

(2)

(   )

(3)

(   )

(4)

(   )

(5)

(   )

(6)

(   )

You're a star. Don't worry if it is difficult, you can do it.

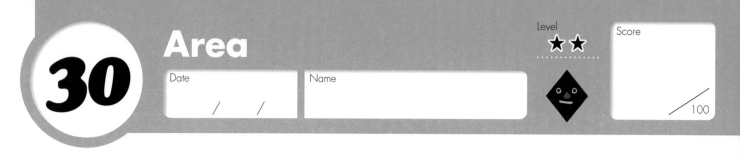

30 Area

Level ★★

Date   /   /

Name

Score   /100

---

1 Find the area of the quadrilateral pictured on the right.

6 points per question

(1) What is the area of triangle ABC?

(   )

(2) What is the area of triangle ACD?

(   )

(3) What is the area of the quadrilateral?

(   )

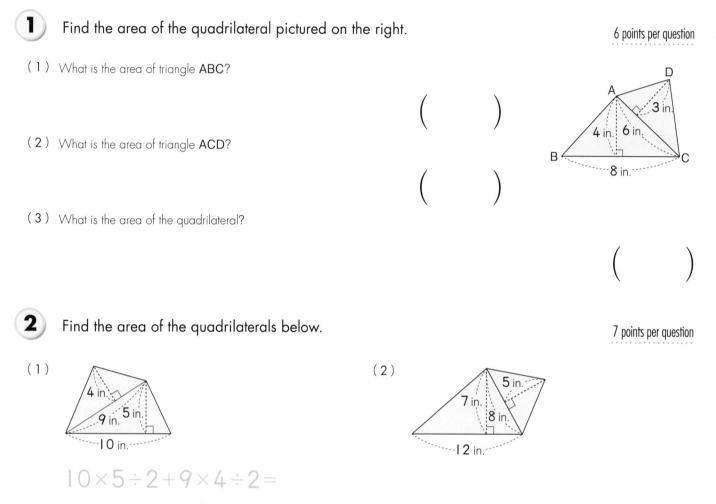

2 Find the area of the quadrilaterals below.

7 points per question

(1)

$10 \times 5 \div 2 + 9 \times 4 \div 2 =$

(   )

(2)

(   )

(3)

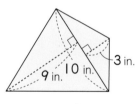

(   )

(4)

(   )

3 Find the area of the shaded parts of the figures below.

6 points per question

(1)

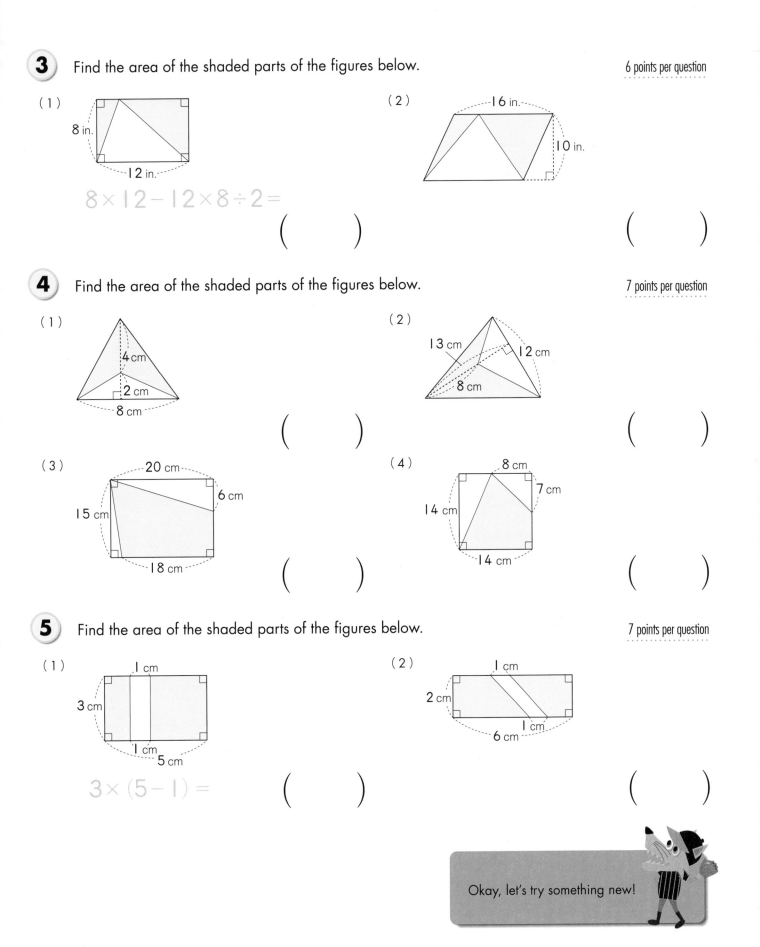

8 in.

12 in.

$8 \times 12 - 12 \times 8 \div 2 =$

(   )

(2)

16 in.

10 in.

(   )

4 Find the area of the shaded parts of the figures below.

7 points per question

(1)

4 cm

2 cm

8 cm

(   )

(2)

13 cm

12 cm

8 cm

(   )

(3)

20 cm

6 cm

15 cm

18 cm

(   )

(4)

8 cm

7 cm

14 cm

14 cm

(   )

5 Find the area of the shaded parts of the figures below.

7 points per question

(1)

1 cm

3 cm

1 cm

5 cm

$3 \times (5 - 1) =$

(   )

(2)

1 cm

2 cm

1 cm

6 cm

(   )

Okay, let's try something new!

Volume

31

Date / / Name

Level ★★★

Score /100

Don't forget!

The volume of a rectangular prism is length times width times height. The volume of a cube is side times side times side.

length · width · height

side · side · side

---

1 Find the volume of the rectangular prisms below.

6 points per question

(1) 2 in. · 1 in. · 3 in.

( )

(2) 5 in. · 3 in. · 4 in.

( )

(3) 6 in. · 12 in. · 8 in.

( )

(4) 3 in. · 3 in. · 3 in.

( )

Don't forget!

The volume of a cube with sides of 1 foot is called 1 cubic foot, and is written 1 ft.³

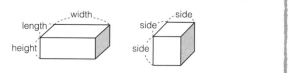

⟨Example⟩ 2 ft. · 3 ft. · 1 ft.

1 ft. · 1 ft. · 1 ft.³

$3 \times 2 \times 1 = 6$
6 ft.³

---

2 Find the volume of the rectangular prisms below. Answer in cubic feet.

6 points per question

(1) 2 ft. · 7 ft. · 5 ft.

( )

(2) 2 ft. · 2 ft. · 2 ft.

( )

---

3 Find the volume of the rectangular prisms below.

8 points per question

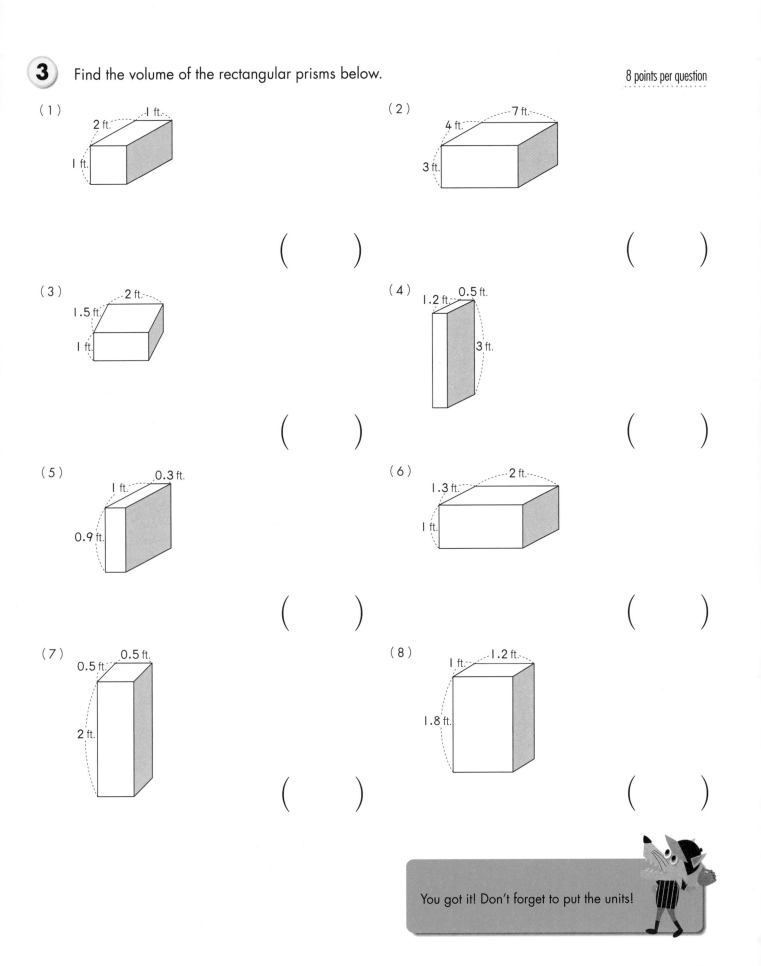

(1)

2 ft.   1 ft.

1 ft.

(   )

(2)

4 ft.   7 ft.

3 ft.

(   )

(3)

2 ft.

1.5 ft.

1 ft.

(   )

(4)

1.2 ft.   0.5 ft.

3 ft.

(   )

(5)

1 ft.   0.3 ft.

0.9 ft.

(   )

(6)

1.3 ft.   2 ft.

1 ft.

(   )

(7)

0.5 ft.   0.5 ft.

2 ft.

(   )

(8)

1 ft.   1.2 ft.

1.8 ft.

(   )

You got it! Don't forget to put the units!

63

Volume

Date / /   Name

Level ★★

Score /100

1 What is the volume of each rectangular prism below?

6 points per question

(1)

3 cm
1 cm
2 cm

( )

(2)

6 cm
2 cm
3 cm

( )

(3)

5 cm
11 cm
9 cm

( )

(4)

4 cm
4 cm
4 cm

( )

Don't forget!

The volume of a cube with sides of 1 m is 1 cubed meter, which is written 1 m³.

1 m
1 m
1 m
1 m³

⟨Example⟩

3 m
4 m
2 m

4 × 3 × 2 = 24
24 m³

---

2 What is the volume of each rectangular prism below?

6 points per question

(1)

1 m
6 m
4 m

( )

(2)

3 m
3 m
3 m

( )

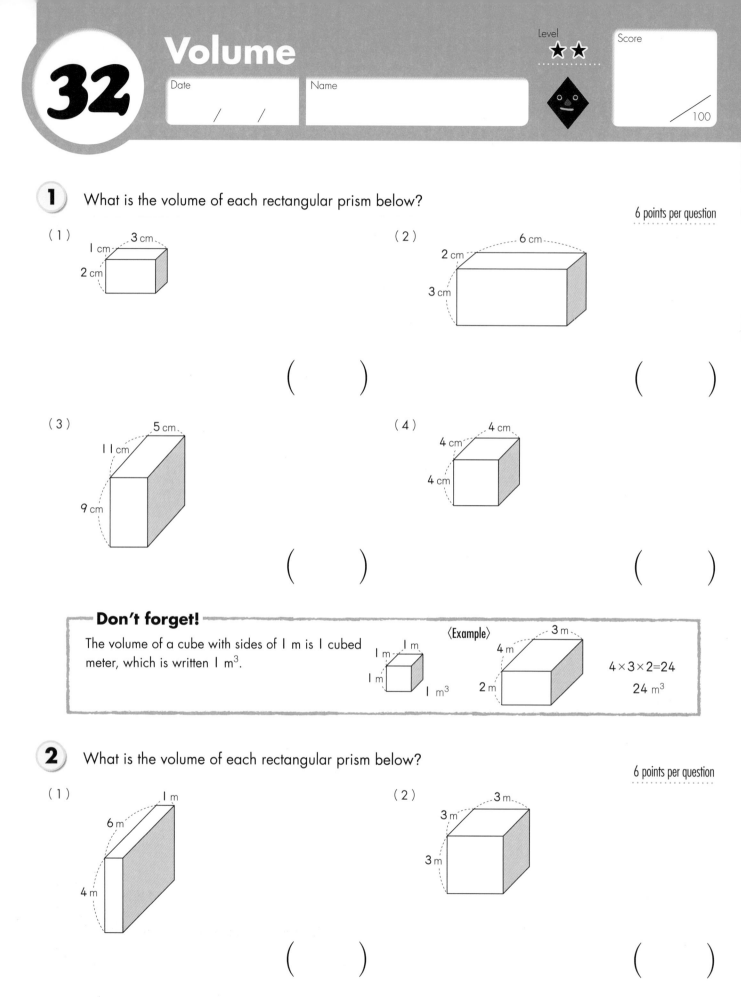

3 What is the volume of each rectangular prism below?   8 points per question

(1)

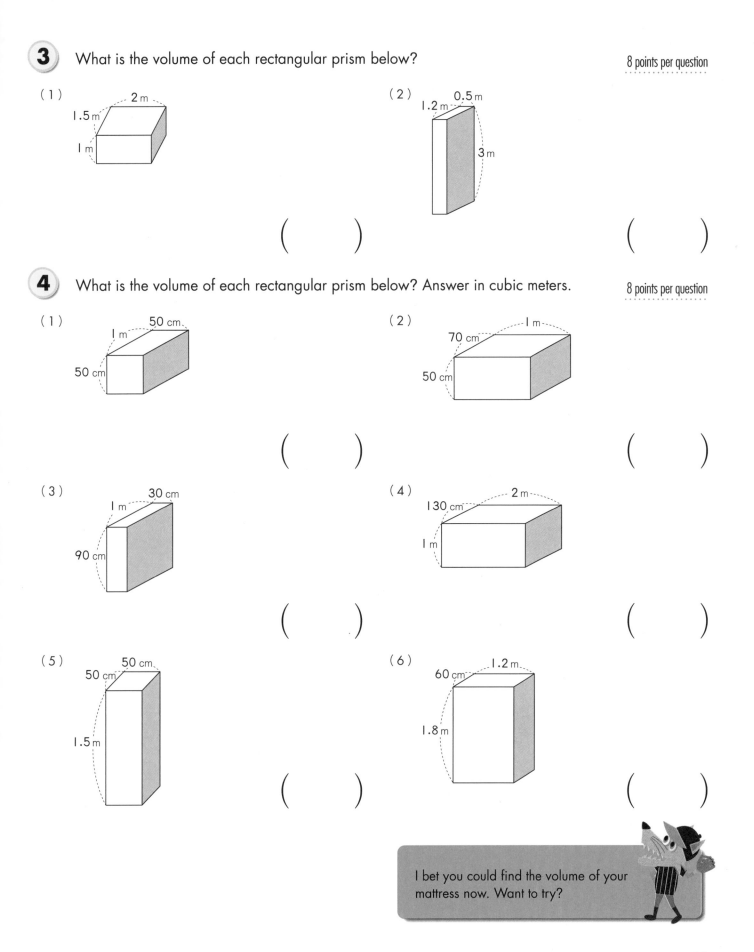

2 m
1.5 m
1 m

(   )

(2)

0.5 m
1.2 m
3 m

(   )

4 What is the volume of each rectangular prism below? Answer in cubic meters.   8 points per question

(1)

50 cm
1 m
50 cm

(   )

(2)

1 m
70 cm
50 cm

(   )

(3)

30 cm
1 m
90 cm

(   )

(4)

2 m
130 cm
1 m

(   )

(5)

50 cm
50 cm
1.5 m

(   )

(6)

1.2 m
60 cm
1.8 m

(   )

I bet you could find the volume of your mattress now. Want to try?

65

Volume

Level ★★

Score /100

Don't forget!

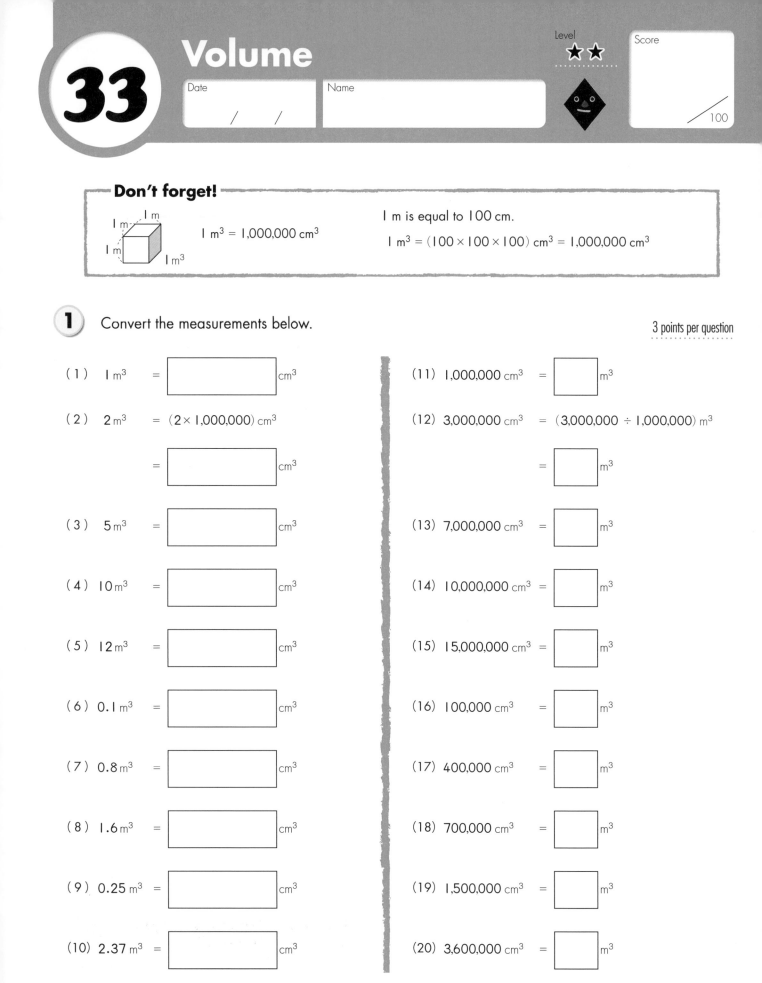

$1 \text{ m}^3 = 1,000,000 \text{ cm}^3$

1 m is equal to 100 cm.

$1 \text{ m}^3 = (100 \times 100 \times 100) \text{ cm}^3 = 1,000,000 \text{ cm}^3$

1 Convert the measurements below.

3 points per question

(1) 1 m^3 = ☐ cm^3

(2) 2 m^3 = $(2 \times 1,000,000) \text{ cm}^3$

= ☐ cm^3

(3) 5 m^3 = ☐ cm^3

(4) 10 m^3 = ☐ cm^3

(5) 12 m^3 = ☐ cm^3

(6) 0.1 m^3 = ☐ cm^3

(7) 0.8 m^3 = ☐ cm^3

(8) 1.6 m^3 = ☐ cm^3

(9) 0.25 m^3 = ☐ cm^3

(10) 2.37 m^3 = ☐ cm^3

(11) $1,000,000 \text{ cm}^3$ = ☐ m^3

(12) $3,000,000 \text{ cm}^3$ = $(3,000,000 \div 1,000,000) \text{ m}^3$

= ☐ m^3

(13) $7,000,000 \text{ cm}^3$ = ☐ m^3

(14) $10,000,000 \text{ cm}^3$ = ☐ m^3

(15) $15,000,000 \text{ cm}^3$ = ☐ m^3

(16) $100,000 \text{ cm}^3$ = ☐ m^3

(17) $400,000 \text{ cm}^3$ = ☐ m^3

(18) $700,000 \text{ cm}^3$ = ☐ m^3

(19) $1,500,000 \text{ cm}^3$ = ☐ m^3

(20) $3,600,000 \text{ cm}^3$ = ☐ m^3

Don't forget!

$1\ L = 1,000\ cm^3$ 　　　　$1\ L = 1,000\ mL$ ➡ $1\ mL = 1\ cm^3$

2 Convert the measurements below.

2 points per question

(1) 1 L = ☐ cm³

(9) 1,000 cm³ = ☐ L

(2) 2 L = (1,000 × 2) cm³

= ☐ cm³

(10) 3,000 cm³ = (3,000 ÷ 1,000) L

= ☐ L

(3) 4 L = ☐ cm³

(11) 7,000 cm³ = ☐ L

(4) 15 L = ☐ cm³

(12) 12,000 cm³ = ☐ L

(5) 0.1 L = ☐ cm³

(13) 100 cm³ = ☐ L

(6) 0.5 L = ☐ cm³

(14) 700 cm³ = ☐ L

(7) 1.2 L = ☐ cm³

(15) 1,800 cm³ = ☐ L

(8) 13.3 L = ☐ cm³

(16) 2,450 cm³ = ☐ L

3 Convert the measurements below.

2 points per question

(1) 1 mL = ☐ cm³

(3) 1 cm³ = ☐ mL

(2) 200 mL = ☐ cm³

(4) 15 cm³ = ☐ mL

Just a little bit more. You can do it!

Volume

Date   /   /

Name

Level
★ ★

Score
/ 100

1 Find the volume of each figure below.

10 points per question

(1)

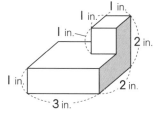

$$\boxed{1 \times 1 \times 1} + \boxed{2 \times 3 \times 1}$$

$= 1 + 6$

$=$ (   )

(2)

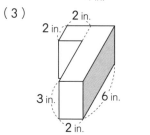

$$\boxed{} + \boxed{}$$

$= \boxed{} + \boxed{}$

$=$ (   )

(3)

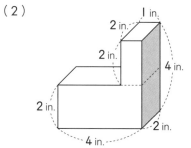

(   )

(4)

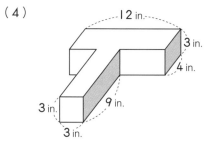

(   )

(5)

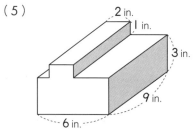

(   )

2 Find the volume of each figure below.

(1)

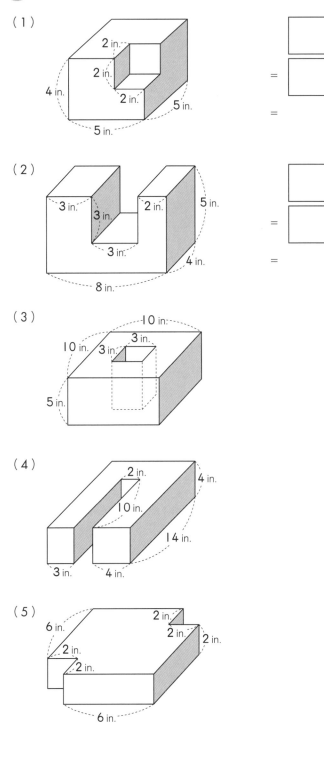

□□□□ − □□□□

= □□□ − □□

= (   )

(2)

□□□□ − □□□□

= □□□ − □□

= (   )

(3)

(   )

(4)

(   )

(5)

(   )

Okay! Let's review what you've learned.

1   You have three cards with the digits 2, 3 and 4 on them.

6 points per question

(1) Write all the even numbers that you can make with the cards.

$\Big($ $\Big)$

(2) Write the biggest odd number that you can make with the cards.

$\big($ $\big)$

2   Write the appropriate numbers below.

5 points per question

(1) The number you get from adding 3 parts of 1, 5 parts of 0.1, and 8 parts of 0.01.

$\Big($ $\Big)$

(2) The number you get from adding 7 parts of 0.1 and 3 parts of 0.001.

$\Big($ $\Big)$

3   Convert each fraction to a decimal and each decimal to a fraction.

5 points per question

(1) $\frac{3}{4}$ =

(2) 0.7 =

(3) $\frac{9}{8}$ =

(4) 2.01 =

4   Circle the larger number in each pair below.

5 points per question

(1) $\left[\ 0.72 \quad \frac{5}{7}\ \right]$

(2) $\left[\ \frac{5}{6} \quad 0.83\ \right]$

5   The line C intersects the parallel lines A and B.

5 points per question

(1) Find angle **a**.

$\Big($ $\Big)$

(2) Find angle **b**.

$\Big($ $\Big)$

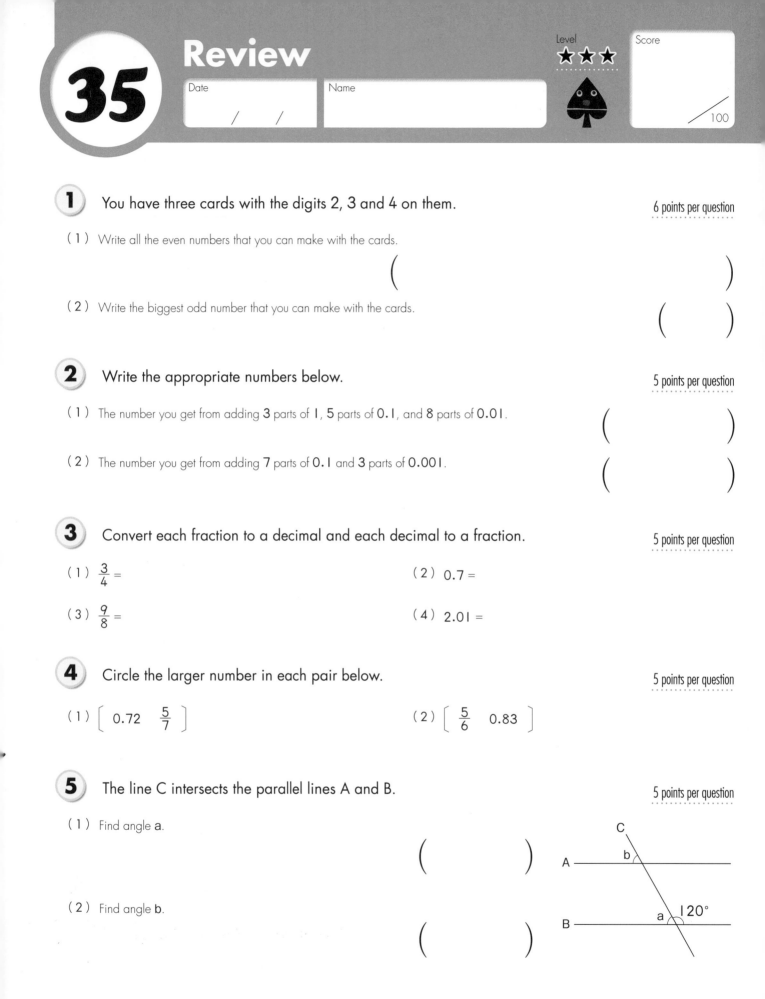

6 The quadrilateral pictured here is a parallelogram.

(1) What is the length of side **AD**?

(   )

(2) Find angle **a**.

(   )

(3) Find angle **b**.

(   )

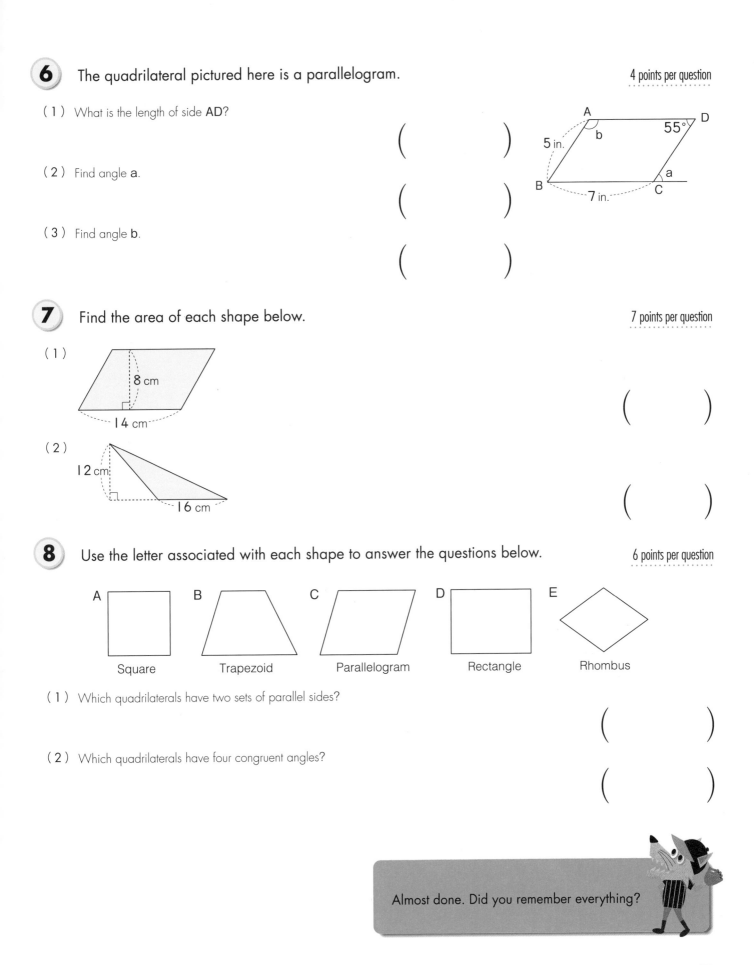

7 Find the area of each shape below.

7 points per question

(1)

8 cm

14 cm

(   )

(2)

12 cm

16 cm

(   )

8 Use the letter associated with each shape to answer the questions below.

6 points per question

A   Square   B   Trapezoid   C   Parallelogram   D   Rectangle   E   Rhombus

(1) Which quadrilaterals have two sets of parallel sides?

(   )

(2) Which quadrilaterals have four congruent angles?

(   )

Almost done. Did you remember everything?

   71

36 **Review**

Level ★★★

Date / /

Name

Score / 100

1 Sort all of the numbers below into even and odd numbers.

12 points for completion

4, 9, 15, 26, 41, 58, 67, 103, 172, 230

Odd ( )

Even ( )

2 Rewrite each division problem below as a fraction.

5 points per question

(1) 3 ÷ 5 =

(2) 9 ÷ 4 =

(3) 8 ÷ 15 =

(4) 12 ÷ 7 =

3 Convert each decimal below into a fraction.

5 points per question

(1) 0.9 =

(2) 0.27 =

(3) 1.7 =

(4) 2.03 =

4 The quadrilateral below is a rhombus.

4 points per question

(1) How long is side **CD**?

( )

(2) Find angle **a**.

( )

(3) How long is the diagonal **AC**?

( )

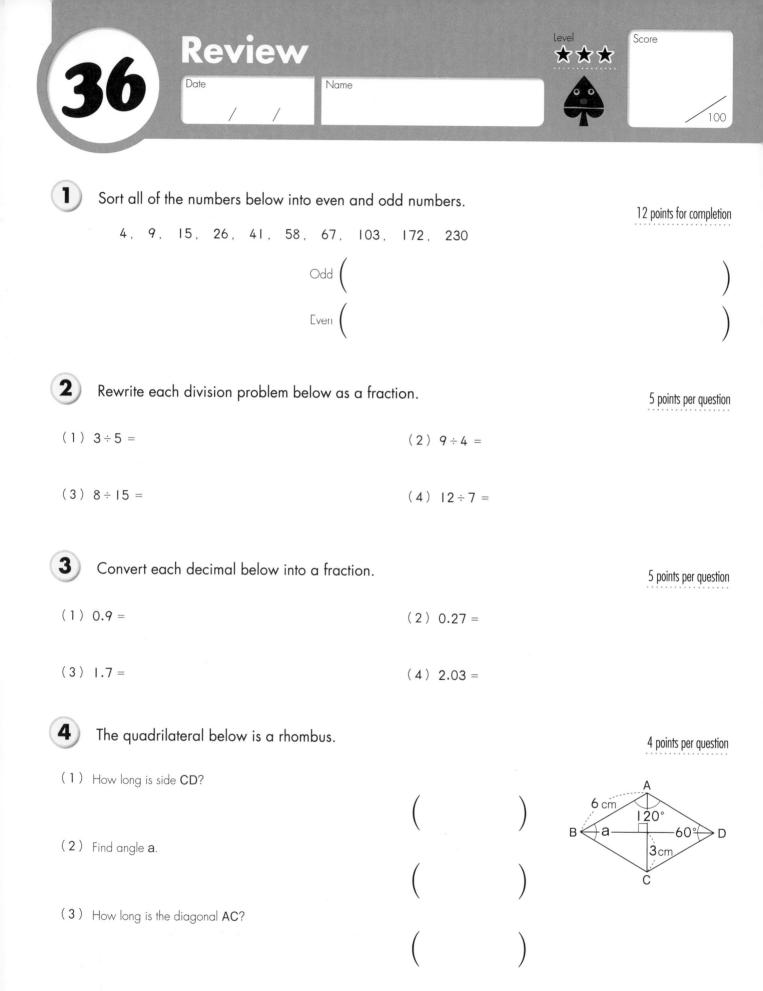

5 Find each angle A below.

6 points per question

(1)

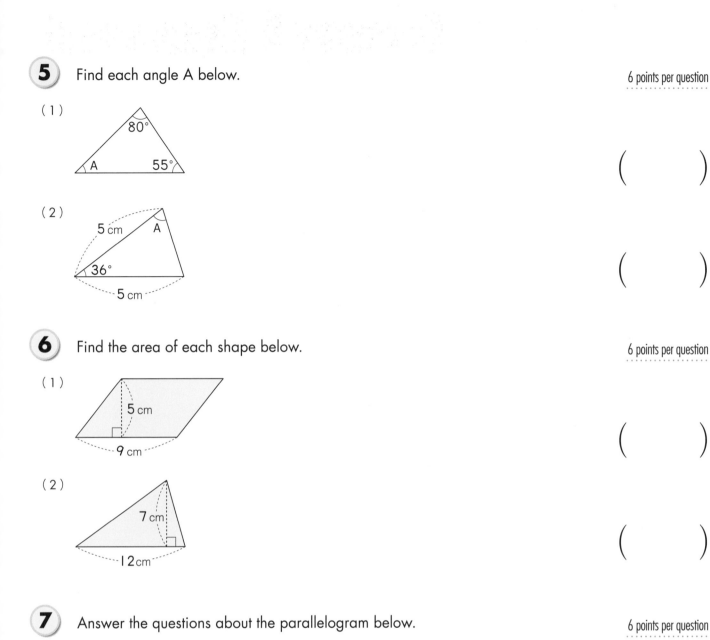

(　　　　)

(2)

5 cm   A

36°

5 cm

(　　　　)

6 Find the area of each shape below.

6 points per question

(1)

5 cm

9 cm

(　　　　)

(2)

7 cm

12 cm

(　　　　)

7 Answer the questions about the parallelogram below.

6 points per question

(1) How long is the line **ED**?

(　　　　)

(2) How long is the line **AE**?

(　　　　)

A   D

7 in.

5 in.

E

B   C

You did it. Congratulations!

(1) Review pp 2, 3

1. 4, 3, 2, 1, 0
2. (1) 32 (2) 5.9 (3) 4.8
3. (1) $1\frac{2}{7}$ (2) 3
4. (1) 3×5=15 **Ans.** 15 in.²
   (2) 6×6=36 **Ans.** 36 in.²
5. (1) 5 cm (2) 5 cm
6. (1) Equilateral triangle (2) Isosceles triangle
7. (1) 105° (2) 60°
8. (1) 16 oz, 2 cups (2) 20 oz, $2\frac{1}{2}$ cups

(2) Review pp 4, 5

1. From 4,450 to 4,549
2. (1) 3 (2) $\frac{1}{8}$ (3) $\frac{5}{5}$, 1
3. (1) 0.7 (2) 5.2 (3) 9.9 (4) 26
4. 8×7−3×5=41 **Ans.** 41 m²
5. (1) $\frac{4}{3}$ (2) $\frac{14}{5}$
6. (1) 3 in. (2) 12 in.
7.
(3 cm) (3 cm) (4 cm)

Advice
Use a ruler to check your answer.

8. (1) 70° (2) 145°
9. (1) 1:30 p.m. (2) 4:20 p.m. (3) 1:15 p.m.

(3) Odd & Even Numbers pp 6, 7

1. (1) 6, 8, 10 (2) 7, 9
2. (1) Odd (2) Even (3) Even (4) Odd (5) Odd
   (6) Odd (7) Odd (8) Even (9) Even (10) Even
   (11) Odd (12) Even
3. (1) 12, 14, 16, 18, 20, 22, 24, 26, 28, 30
   (2) 11, 13, 15, 17, 19, 21, 23, 25, 27, 29
4. 46, 68, 96, 104, 126, 156, 160, 196
5. 43, 69, 91, 117, 145, 151, 179, 199
6. (Even) 36, 126, 180, 252, 300, 78, 98, 112, 136,
   284, 176, 184
   (Odd) 43, 193, 83, 67, 145, 321, 267, 365, 243

7. (1) 132, 312 (2) 123, 213, 231, 321
8. 423

(4) Decimals pp 8, 9

1. (1) 7 (2) 9 (3) 10 (4) 11
2. (1) 4 (2) 8 (3) 2 (4) 3 (5) 10 (6) 11
3. (1) 6 (2) 5 (3) 2 (4) 3 (5) 8 (6) 10
4. (1) 0.3 (2) 0.7 (3) 1.5 (4) 4.8
5. (1) 0.05 (2) 0.09 (3) 0.15 (4) 1.25
6. (1) 0.004 (2) 0.018 (3) 0.392 (4) 1.549
7. (1) 3 (2) 4 (3) 5 (4) 6
8. (1) 3, 5, 1, 2 (2) 0, 2, 7, 5 (3) 0, 5, 0, 8
9. (1) 2.35 (2) 3.417 (3) 2.704 (4) 0.864

(5) Decimals pp 10, 11

1. (1) (From the left) 0.01, 0.08, 0.12, 0.19
   (2) 0.04, 0.15, 0.27, 0.32
   (3) 0.001, 0.006, 0.012, 0.018
   (4) 0.002, 0.014, 0.025, 0.033
2.
C A B D
0 0.1 0.2
3. (1) 0.13 (2) 0.27 (3) 0.05
   (4) 0.1 (5) 3.42 (6) 0.16
4. (1) 3.47 (2) 4.23 (3) 5.47 (4) 2.138
5. (1) 1.11, 1.04, 1.001
   (2) 9.111, 9.11, 9.01, 9.001

(6) Decimals pp 12, 13

1. (1) 20, 20 (2) 2, 2 (3) 0.2, 0.2
2. (1) 20, 20 (2) 2, 2
3. (1) 3 (2) 0.3 (3) 9 (4) 0.5 (5) 4
   (6) 0.4 (7) 7 (8) 0.8
4. (1) 2, 2 (2) 0.2, 0.2 (3) 0.02, 0.02
5. (1) 0.02, 0.02 (2) 0.002, 0.002
6. (1) 0.4 (2) 0.04 (3) 0.7 (4) 0.06
   (5) 0.03 (6) 0.003 (7) 0.08 (8) 0.005

7 Decimals
pp 14,15

1 (1) 32.6, 1   (2) 326, 2

2 (1) 4   (2) 42   (3) 25.7   (4) 463.2   (5) 70
   (6) 32   (7) 247   (8) 1,824   (9) 306   (10) 4,105

3 (1) 10   (2) 100

4 (1) 3.26, 1   (2) 0.326, 2

5 (1) 3.2   (2) 0.24   (3) 0.257   (4) 4.87
   (5) 4.63   (6) 2.3   (7) 0.158   (8) 0.075

6 (1) $\frac{1}{10}$ (2) $\frac{1}{100}$

7   3,265, 32,650, 32.65, 3.265

8 Decimals & Measurements
pp 16,17

1 (1) 2   (2) 3   (3) 0.3   (4) 0.03   (5) 0.05
   (6) 0.35   (7) 0.46   (8) 1.35   (9) 1.28

2 (1) 0.4   (2) 0.04   (3) 0.08   (4) 0.17   (5) 0.65
   (6) 1.25   (7) 1.1   (8) 1.01   (9) 2.45   (10) 3.08

3 (1) 3   (2) 0.3   (3) 0.03   (4) 0.003
   (5) 0.354   (6) 0.065   (7) 1.45   (8) 2.387

4 (1) 0.6   (2) 0.15   (3) 0.04   (4) 0.008
   (5) 0.275   (6) 0.036   (7) 1.48   (8) 2.306

5 (1) 0.5   (2) 0.03   (3) 0.725
   (4) 0.064   (5) 1.82   (6) 2.905

9 Decimals & Fractions
pp 18,19

1   B : $\frac{1}{3}$ C : (From the left) $\frac{2}{4}$, $\frac{3}{4}$ D : $\frac{2}{5}$ E : $\frac{3}{6}$, $\frac{5}{6}$
F : $\frac{2}{7}$, $\frac{4}{7}$ G : $\frac{1}{8}$, $\frac{4}{8}$, $\frac{7}{8}$ H : $\frac{2}{9}$, $\frac{5}{9}$ I : $\frac{1}{10}$, $\frac{9}{10}$

2 (1) $\frac{2}{4}$, $\frac{3}{6}$, $\frac{4}{8}$, $\frac{5}{10}$ (2) $\frac{2}{6}$, $\frac{3}{9}$ (3) $\frac{2}{3}$, $\frac{4}{6}$

3 (1) $\frac{3}{4}$ (2) $\frac{2}{3}$

4 (1) $\frac{4}{5}$, $\frac{3}{5}$, $\frac{2}{5}$, $\frac{1}{5}$ (2) $\frac{3}{4}$, $\frac{3}{7}$, $\frac{3}{8}$, $\frac{3}{9}$

5 (1) $\frac{2}{8}$, $\frac{5}{8}$, $\frac{9}{8}$, $\frac{11}{8}$ (2) $\frac{4}{9}$, $\frac{4}{7}$, $\frac{4}{5}$, $\frac{4}{3}$ (3) $\frac{3}{6}$, $\frac{5}{6}$, $\frac{6}{6}$, $\frac{7}{6}$

10 Decimals & Fractions
pp 20,21

1 (1) $\frac{3}{4}$ (2) $\frac{5}{7}$ (3) $\frac{1}{3}$ (4) $\frac{7}{11}$ (5) $\frac{4}{9}$
   (6) $\frac{9}{10}$ (7) $\frac{7}{8}$ (8) $\frac{3}{7}$ (9) $\frac{5}{12}$ (10) $\frac{9}{14}$
   (11) $\frac{5}{4}$ (12) $\frac{7}{3}$ (13) $\frac{10}{7}$ (14) $\frac{9}{4}$ (15) $\frac{11}{6}$
   (16) $\frac{6}{5}$ (17) $\frac{12}{7}$ (18) $\frac{15}{8}$

---

2 (1) 5   (2) 1   (3) 7   (4) 9

3 (1) 0.4   (2) 0.6   (3) 0.8   (4) 1.2
   (5) 0.75   (6) 1.25   (7) 1.5   (8) 2.5
   (9) 0.875   (10) 1.125   (11) 0.1   (12) 1.7
   (13) 0.25   (14) 0.125   (15) 0.0625   (16) 0.3125

4 (1) $\frac{3}{10}$ (2) $\frac{7}{10}$ (3) $\frac{9}{10}$ (4) $\frac{4}{10}$ (5) $\frac{5}{10}$
   (6) $\frac{11}{10}$ (7) $\frac{13}{10}$ (8) $\frac{18}{10}$ (9) $\frac{27}{10}$ (10) $\frac{39}{10}$
   (11) $\frac{3}{100}$ (12) $\frac{7}{100}$ (13) $\frac{9}{100}$ (14) $\frac{8}{100}$ (15) $\frac{13}{100}$
   (16) $\frac{27}{100}$ (17) $\frac{35}{100}$ (18) $\frac{113}{100}$ (19) $\frac{223}{100}$ (20) $\frac{107}{100}$

5 (1) $\frac{1}{10}$ (2) $\frac{15}{10}$ (3) $\frac{71}{100}$ (4) $\frac{249}{100}$ (5) $\frac{101}{100}$
   (6) 0.9   (7) 0.8   (8) 1.5   (9) 0.35   (10) 0.37

6 (1) $\frac{1}{2}$ (2) $\frac{3}{10}$ (3) $\frac{8}{5}$ (4) $\frac{36}{100}$

7 $\frac{9}{5}$, 1.6, $1\frac{1}{2}$, $\frac{3}{4}$, 0.7

11 Perpendicular & Parallel Lines
pp 22,23

1   a, d

2   B, C

3   B, D, F

4   A and C,   B and E

5   A and E,   B and D,   C and F

12 Perpendicular & Parallel Lines
pp 24,25

1 (1) b, c   (2) e, f   (3) h, i

2   (a, c, e, g), (b, d, f, h)

3 (1) 5 in.   (2) 70°
   (3) 180−70=110   **Ans.** 110°
   (4) 70+110=180   **Ans.** 180°

4 (1) c, e
   (2) 180−120=60   **Ans.** 60°
   (3) d, f
   (4) 60+120=180   **Ans.** 180°

5 (1) a, d, e, f
   (2) 180−65=115   **Ans.** 115°
   (3) c

6 (1)
   (2) (retangle) ab and dc,   ad and bc
       (square) ef and hg,   eh and fg

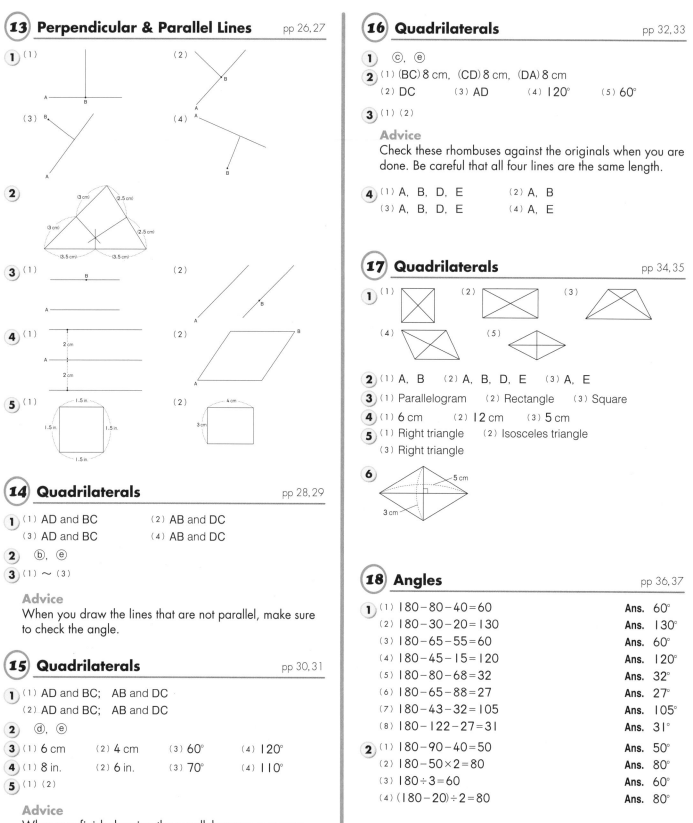

13 Perpendicular & Parallel Lines   pp 26, 27

1 (1)   (2)

(3)   (4)

2

3 (1)   (2)

4 (1)   (2)

5 (1)   (2)

14 Quadrilaterals   pp 28, 29

1 (1) AD and BC   (2) AB and DC

(3) AD and BC   (4) AB and DC

2 ⓑ, ⓔ

3 (1) ~ (3)

Advice
When you draw the lines that are not parallel, make sure to check the angle.

15 Quadrilaterals   pp 30, 31

1 (1) AD and BC;   AB and DC

(2) AD and BC;   AB and DC

2 ⓓ, ⓔ

3 (1) 6 cm   (2) 4 cm   (3) 60°   (4) 120°

4 (1) 8 in.   (2) 6 in.   (3) 70°   (4) 110°

5 (1) (2)

Advice
When you finish drawing the parallelograms, compare them to the originals.

16 Quadrilaterals   pp 32, 33

1 ⓒ, ⓔ

2 (1) (BC) 8 cm, (CD) 8 cm, (DA) 8 cm

(2) DC   (3) AD   (4) 120°   (5) 60°

3 (1) (2)

Advice
Check these rhombuses against the originals when you are done. Be careful that all four lines are the same length.

4 (1) A, B, D, E   (2) A, B

(3) A, B, D, E   (4) A, E

17 Quadrilaterals   pp 34, 35

1 (1)   (2)   (3)

(4)   (5)

2 (1) A, B   (2) A, B, D, E   (3) A, E

3 (1) Parallelogram   (2) Rectangle   (3) Square

4 (1) 6 cm   (2) 12 cm   (3) 5 cm

5 (1) Right triangle   (2) Isosceles triangle

(3) Right triangle

6

18 Angles   pp 36, 37

1 (1) 180−80−40=60   **Ans.** 60°

(2) 180−30−20=130   **Ans.** 130°

(3) 180−65−55=60   **Ans.** 60°

(4) 180−45−15=120   **Ans.** 120°

(5) 180−80−68=32   **Ans.** 32°

(6) 180−65−88=27   **Ans.** 27°

(7) 180−43−32=105   **Ans.** 105°

(8) 180−122−27=31   **Ans.** 31°

2 (1) 180−90−40=50   **Ans.** 50°

(2) 180−50×2=80   **Ans.** 80°

(3) 180÷3=60   **Ans.** 60°

(4) (180−20)÷2=80   **Ans.** 80°

3 (1) 180−70=110, 180−30−110=40   **Ans.** 40°

(2) 180−85−65=30, 180−30=150   **Ans.** 150°

(3) 180−110=70, 180−105=75,
180−70−75=35   **Ans.** 35°

(4) 180−95=85, 180−40−85=55,
180−55=125   **Ans.** 125°

(5) 180−90=90, 180−125=55,
180−90−55=35, 180−35=145   **Ans.** 145°

(6) 180−150=30, 180−95=85,
180−30−85=65, 180−65=115   **Ans.** 115°

(19) Angles
pp 38,39

1 360°

Advice
The sum of all the angles is the sum of the angles of two
triangles.   180×2=360

2 (1) 360−95−115−80=70   **Ans.** 70°

(2) 360−90−95−95=80   **Ans.** 80°

(3) 360−80−85−90=105   **Ans.** 105°

(4) 360−150−46−94=70   **Ans.** 70°

(5) 180−125=55,
360−85−120−55=100   **Ans.** 100°

(6) 180−89=91, 180−88=92,
360−91−92−45=132   **Ans.** 132°

3 540°

Advice
The sum of all the angles is the sum of the angles of three
triangles.   180×3=540

4 (1) 540−95−120−90−100=135   **Ans.** 135°

(2) 540−100−123−86−87=144   **Ans.** 144°

5 720°

Advice
The sum of all the angles is the sum of the angles of four
triangles.   180×4=720

6 (1) 720−100−155−145−110−95=115   **Ans.** 115°

(2) 720−85−150−134−116−145=90   **Ans.** 90°

(20) Polygons
pp 40,41

1 (1) Regular hexagon   (2) Regular pentagon

(3) Regular octagon   (4) Regular decagon

2 ⓐ ×   ⓑ ✓   ⓒ ✓   ⓓ ✓   ⓔ ✓   ⓕ ×   ⓖ ×   ⓗ ×

3 (1) 360÷6=60   **Ans.** 60°

(2) 360÷8=45   **Ans.** 45°

(3) 360÷5=72   **Ans.** 72°

(4) 360÷10=36   **Ans.** 36°

4 (1) 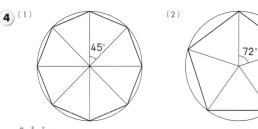 (2)

Advice
Make sure to check the example!

(21) Congruent Figures
pp 42,43

1 F, H

2 A and H,   C and G,   D and K,   F and I

3 (1) D   (2) E   (3) F

(4) DE   (5) EF   (6) FD

(7) angle D   (8) angle E   (9) angle F

4 (1) B   (2) GF   (3) angle E

(4) 2 cm   (5) 3 cm   (6) 60°

(22) Congruent Figures
pp 44,45

1 (1) DC   (2) AD   (3) AC   (4) d

(5) e   (6) f   (7) congruent

2 Triangle ABD and Triangle CDB

3 (1) Triangle CDO   (2) Triangle DAO

4 (1) Triangle CDB,   Triangle BCA,   Triangle DAC

(2) Triangle CDO   (3) Triangle DAO

5 (1) Triangle CDB

(2) Triangle ADO and Triangle CBO and Triangle CDO

6 (1) Triangle BCO and Triangle CDO and Triangle DAO

(2) Triangle BCA and Triangle CDB and Triangle DAC

(23) Axial Symmetry
pp 46,47

1 A, B, C

2 A, C, D

3 A, C, D, E, F, H

4 (1)   (2)

(3)   (4)

   77

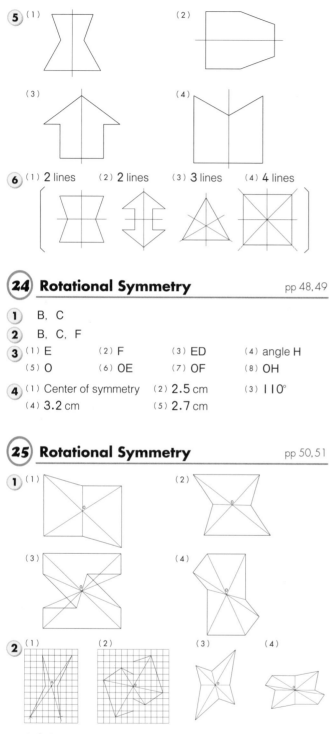

5 (1) (2)

(3) (4)

6 (1) **2** lines (2) **2** lines (3) **3** lines (4) **4** lines

24 **Rotational Symmetry** pp 48,49

1 B, C

2 B, C, F

3 (1) E (2) F (3) ED (4) angle H
(5) O (6) OE (7) OF (8) OH

4 (1) Center of symmetry (2) **2.5** cm (3) **110°**
(4) **3.2** cm (5) **2.7** cm

25 **Rotational Symmetry** pp 50,51

1 (1) (2)

(3) (4)

2 (1) (2) (3) (4)

Advice
Use these steps to draw your figures.
(1) Draw lines that start at the existing vertices and go to the center of symmetry.
(2) Measure the length from each vertex to the center of symmetry along the lines you have drawn.
(3) Extending the lines you have drawn, travel that same distance from the center of symmetry to the new vertex.

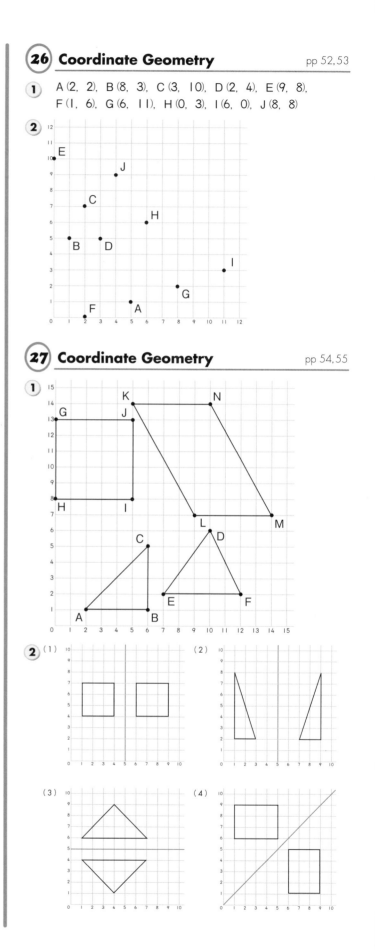

26 **Coordinate Geometry** pp 52,53

1 A (2, 2), B (8, 3), C (3, 10), D (2, 4), E (9, 8), F (1, 6), G (6, 11), H (0, 3), I (6, 0), J (8, 8)

2

27 **Coordinate Geometry** pp 54,55

1

2 (1) (2) (3) (4)

28 Area
pp 56,57

1
(1) $4 \times 3 = 12$ — Ans. 12 in.²
(2) $5 \times 7 = 35$ — Ans. 35 in.²
(3) $6 \times 8 = 48$ — Ans. 48 in.²
(4) $10 \times 7 = 70$ — Ans. 70 in.²
(5) $8 \times 9 = 72$ — Ans. 72 in.²
(6) $14 \times 5 = 70$ — Ans. 70 in.²
(7) $16 \times 4 = 64$ — Ans. 64 in.²
(8) $7 \times 11 = 77$ — Ans. 77 in.²

2
(1) 12 cm
(2) $5 \times 12 = 60$ — Ans. 60 cm²

3
(1) $4 \times 8 = 32$ — Ans. 32 cm²
(2) $7 \times 12 = 84$ — Ans. 84 cm²
(3) $6 \times 12 = 72$ — Ans. 72 cm²
(4) $5 \times 11 = 55$ — Ans. 55 cm²
(5) $5 \times 15 = 75$ — Ans. 75 cm²
(6) $7 \times 18 = 126$ — Ans. 126 cm²

29 Area
pp 58,59

1
(1) $4 \times 3 \div 2 = 6$ — Ans. 6 in.²
(2) $5 \times 6 \div 2 = 15$ — Ans. 15 in.²
(3) $4 \times 8 \div 2 = 16$ — Ans. 16 in.²
(4) $7 \times 7 \div 2 = 24.5$ — Ans. 24.5 in.²
(5) $10 \times 5 \div 2 = 25$ — Ans. 25 in.²
(6) $12 \times 7 \div 2 = 42$ — Ans. 42 in.²
(7) $16 \times 7 \div 2 = 56$ — Ans. 56 in.²
(8) $11 \times 5 \div 2 = 27.5$ — Ans. 27.5 in.²

2
(1) 14 cm
(2) $6 \times 14 \div 2 = 42$ — Ans. 42 cm²

3
(1) $5 \times 8 \div 2 = 20$ — Ans. 20 cm²
(2) $6 \times 7 \div 2 = 21$ — Ans. 21 cm²
(3) $8 \times 12 \div 2 = 48$ — Ans. 48 cm²
(4) $6 \times 15 \div 2 = 45$ — Ans. 45 cm²
(5) $4 \times 9 \div 2 = 18$ — Ans. 18 cm²
(6) $5 \times 11 \div 2 = 27.5$ — Ans. 27.5 cm²

30 Area
pp 60,61

1
(1) $8 \times 4 \div 2 = 16$ — Ans. 16 in.²
(2) $6 \times 3 \div 2 = 9$ — Ans. 9 in.²
(3) $16 + 9 = 25$ — Ans. 25 in.²

2
(1) $10 \times 5 \div 2 + 9 \times 5 \div 2 = 43$ — Ans. 43 in.²
(2) $8 \times 5 \div 2 + 12 \times 7 \div 2 = 62$ — Ans. 62 in.²
(3) $10 \times 3 \div 2 + 10 \times 9 \div 2 = 60$ — Ans. 60 in.²
(4) $15 \times 5 \div 2 + 15 \times 8 \div 2 = 97.5$ — Ans. 97.5 in.²

3
(1) $8 \times 12 - 12 \times 8 \div 2 = 48$ — Ans. 48 in.²
(2) $16 \times 10 - 16 \times 10 \div 2 = 80$ — Ans. 80 in.²

4
(1) $8 \times (4+2) \div 2 - 8 \times 2 \div 2 = 16$ — Ans. 16 cm²
(2) $12 \times 13 \div 2 - 12 \times (13-8) \div 2 = 48$ — Ans. 48 cm²
(3) $15 \times 20 - (20-18) \times 15 \div 2 - 6 \times 20 \div 2 = 225$ — Ans. 225 cm²
(4) $14 \times 14 - (14-8) \times 14 \div 2 - 8 \times 7 \div 2 = 126$ — Ans. 126 cm²

5
(1) $3 \times (5-1) = 12$ — Ans. 12 cm²
(2) $2 \times (6-1) = 10$ — Ans. 10 cm²

31 Volume
pp 62,63

1
(1) $1 \times 2 \times 3 = 6$ — Ans. 6 in.³
(2) $3 \times 5 \times 4 = 60$ — Ans. 60 in.³
(3) $12 \times 6 \times 8 = 576$ — Ans. 576 in.³
(4) $3 \times 3 \times 3 = 27$ — Ans. 27 in.³

2
(1) $7 \times 2 \times 5 = 70$ — Ans. 70 ft.³
(2) $2 \times 2 \times 2 = 8$ — Ans. 8 ft.³

3
(1) $2 \times 1 \times 1 = 2$ — Ans. 2 ft.³
(2) $4 \times 7 \times 3 = 84$ — Ans. 84 ft.³
(3) $1.5 \times 2 \times 1 = 3$ — Ans. 3 ft.³
(4) $1.2 \times 0.5 \times 3 = 1.8$ — Ans. 1.8 ft.³
(5) $1 \times 0.3 \times 0.9 = 0.27$ — Ans. 0.27 ft.³
(6) $1.3 \times 2 \times 1 = 2.6$ — Ans. 2.6 ft.³
(7) $0.5 \times 0.5 \times 2 = 0.5$ — Ans. 0.5 ft.³
(8) $1 \times 1.2 \times 1.8 = 2.16$ — Ans. 2.16 ft.³

32 Volume
pp 64,65

1
(1) $1 \times 3 \times 2 = 6$ — Ans. 6 cm³
(2) $2 \times 6 \times 3 = 36$ — Ans. 36 cm³
(3) $11 \times 5 \times 9 = 495$ — Ans. 495 cm³
(4) $4 \times 4 \times 4 = 64$ — Ans. 64 cm³

2
(1) $6 \times 1 \times 4 = 24$ — Ans. 24 m³
(2) $3 \times 3 \times 3 = 27$ — Ans. 27 m³

3
(1) $1.5 \times 2 \times 1 = 3$ — Ans. 3 m³
(2) $1.2 \times 0.5 \times 3 = 1.8$ — Ans. 1.8 m³

4
(1) $1 \times 0.5 \times 0.5 = 0.25$ — Ans. 0.25 m³
(2) $0.7 \times 1 \times 0.5 = 0.35$ — Ans. 0.35 m³
(3) $1 \times 0.3 \times 0.9 = 0.27$ — Ans. 0.27 m³
(4) $1.3 \times 2 \times 1 = 2.6$ — Ans. 2.6 m³
(5) $0.5 \times 0.5 \times 1.5 = 0.375$ — Ans. 0.375 m³
(6) $0.6 \times 1.2 \times 1.8 = 1.296$ — Ans. 1.296 m³

33 Volume
pp 66,67

1
(1) 1,000,000
(2) 2,000,000
(3) 5,000,000
(4) 10,000,000
(5) 12,000,000
(6) 100,000
(7) 800,000
(8) 1,600,000
(9) 250,000
(10) 2,370,000
(11) 1
(12) 3
(13) 7
(14) 10
(15) 15
(16) 0.1
(17) 0.4
(18) 0.7
(19) 1.5
(20) 3.6

2
(1) 1,000
(2) 2,000
(3) 4,000
(4) 15,000
(5) 100
(6) 500
(7) 1,200
(8) 13,300
(9) 1
(10) 3
(11) 7
(12) 12
(13) 0.1
(14) 0.7
(15) 1.8
(16) 2.45

3
(1) 1
(2) 200
(3) 1
(4) 15

34 Volume
pp 68,69

1
(1) $1 \times 1 \times 1 + 2 \times 3 \times 1$
$= 1 + 6$
$= 7$ **Ans.** 7 in.³

(2) $2 \times 1 \times 2 + 2 \times 4 \times 2$
$= 4 + 16$
$= 20$ **Ans.** 20 in.³

(3) $2 \times 2 \times 3 + 2 \times 6 \times 3$
$= 12 + 36$
$= 48$ **Ans.** 48 in.³

(4) $3 \times 9 \times 3 + 4 \times 12 \times 3$
$= 81 + 144$
$= 225$ **Ans.** 225 in.³

(5) $9 \times 2 \times 1 + 9 \times 6 \times 3$
$= 18 + 162$
$= 180$ **Ans.** 180 in.³

2
(1) $5 \times 5 \times 4 - 2 \times 2 \times 2$
$= 100 - 8$
$= 92$ **Ans.** 92 in.³

(2) $4 \times 8 \times 5 - 4 \times 3 \times 3$
$= 160 - 36$
$= 124$ **Ans.** 124 in.³

(3) $10 \times 10 \times 5 - 3 \times 3 \times 5$
$= 500 - 45$
$= 455$ **Ans.** 455 in.³

(4) $14 \times (3 + 2 + 4) \times 4 - 10 \times 2 \times 4$
$= 504 - 80$
$= 424$ **Ans.** 424 in.³

(5) $(6 + 2) \times (6 + 2) \times 2 - 2 \times 2 \times 2 \times 2$
$= 128 - 16$
$= 112$ **Ans.** 112 in.³

35 Review
pp 70,71

1
(1) 234, 324, 342, 432
(2) 423

2
(1) 3.58
(2) 0.703

3
(1) 0.75
(2) $\frac{7}{10}$
(3) 1.125
(4) $\frac{201}{100}$

4
(1) 0.72
(2) $\frac{5}{6}$

5
(1) $180 - 120 = 60$ **Ans.** 60°
(2) 60°

6
(1) 7 in.
(2) 55°
(3) 125°

7
(1) $14 \times 8 = 112$ **Ans.** 112 cm²
(2) $16 \times 12 \div 2 = 96$ **Ans.** 96 cm²

8
(1) A, C, D, E
(2) A, D

36 Review
pp 72,73

1
Odd (9, 15, 41, 67, 103)
Even (4, 26, 58, 172, 230)

2
(1) $\frac{3}{5}$
(2) $\frac{9}{4}$
(3) $\frac{8}{15}$
(4) $\frac{12}{7}$

3
(1) $\frac{9}{10}$
(2) $\frac{27}{100}$
(3) $\frac{17}{10}$
(4) $\frac{203}{100}$

4
(1) 6 cm
(2) 60°
(3) 6 cm

5
(1) $180 - 80 - 55 = 45$ **Ans.** 45°
(2) $(180 - 36) \div 2 = 72$ **Ans.** 72°

6
(1) $9 \times 5 = 45$ **Ans.** 45 cm²
(2) $12 \times 7 \div 2 = 42$ **Ans.** 42 cm²

7
(1) 5 in.
(2) 3.5 in. $\left(\frac{7}{2} \text{ in.}\right)$